Advances in Volcanology

An Official Book Series of the International Association of Volcanology and Chemistry of the Earth's Interior

Series Editor

Karoly Nemeth, Institute of Natural Resources, Massey University, Palmerston North, New Zealand

Advances in Volcanology is an official book series of the International Association of Volcanology and Chemistry of the Earth's Interior (IAVCEI). The aim of the book series is to publish scientific monographs on a varied array of topics or themes in volcanology. The series aims at building a varied library of reference works, by describing the current understanding of selected themes such as certain types of volcanism or regional aspects. The books in the series are prepared by leading experts actively working in the relevant field. The Advances in Volcanology series contains single and multi-authored books as well as edited volumes. The Series Editor, Dr. Karoly Nemeth (Massey University, New Zealand), is currently accepting proposals and a proposal document can be obtained from the Publisher, Dr. Annett Buettner (annett.buettner@springer.com).

More information about this series at http://www.springer.com/series/11157

Diego Perugini

The Mixing of Magmas

Field Evidence, Numerical Models, Experiments

Springer

Diego Perugini
Department of Physics and Geology
University of Perugia
Perugia, Italy

ISSN 2364-3277 ISSN 2364-3285 (electronic)
Advances in Volcanology
ISBN 978-3-030-81810-4 ISBN 978-3-030-81811-1 (eBook)
https://doi.org/10.1007/978-3-030-81811-1

This Springer imprint is published by the registered company Springer Nature Switzerland AG
The registered company address is: Gewerbestrasse 11, 6330 Cham, Switzerland

To my wife Laura, who was brave enough to believe in the impossible.

To my son Lorenzo. In the vortexes of chaos, it's a privilege for me to travel next to your trajectory.

To Giulia, who will never be a physicist. Listen to your heart and follow your dreams.

To my parents, Umbra and Nello.

Preface

We adore chaos because we love to produce order.

M. C. Escher

Since the beginning of what we call today Science, researchers have attempted to describe and understand Nature by building models. However, the scientific practice is characterized by a great diversity. It assumes different forms in different disciplines, in different historical periods, according to research schools and individual scientists. To date, a unitary model has not yet been devised accounting for the scientific practice in all its variety and complexity. Until such a model will exist, the best way to describe and understand Nature is to develop partial models, each accounting reasonably for its various aspects. This is what has also happened in the branch of petrology that deals with the process of magma mixing. Although there is a vast literature on this subject, most of the works are focused on the study of particular cases and the results are rarely used for general and unitary interpretations. This may be due to various causes and, among these, probably the most relevant one is that magma mixing can occur in a large number of forms, very variable from case to case, which can be complex to include in a single general conceptual framework. Although, therefore, it may appear legitimate to study this natural process using partial models, the most serious risk is to create a scientific field that is separate from the other disciplines, which could not only benefit from the results of this type of research, but, above all, may provide fundamental indications for a better understanding of the process itself. I refer first of all to those scientific disciplines, such as physics and mathematics, which are often referred to as "exact sciences", and therefore considered barely useful in the study of Nature in which precision and reproducibility are rare qualities. But there is also another risk. A compartmentalized study inevitably causes the marginalization of the importance of a natural process that risks to be considered as a deviation of a natural system, which, therefore, represents the exception rather than the rule. If at the beginning the process attracts the attention of researchers, it is inevitably destined to be relegated to the "geological zoo of bizarre cases" and sooner or later it will be forgotten. This happens because, although that particular case belongs to a much wider class of processes having a single common thread, it lacks in fact of a general fundamental theory that supports it coherently. The main issue is therefore to make efforts in order to generalize the results from the studies carried out on magma mixing with the aim of constructing a common basis to the different occurrences. A possible approach to this problem may come from the application of theories and techniques that allowed to establish universal characters in many scientific fields: Chaos Theory and Fractal Geometry. This is what I attempted to do in the last twenty-five years of research on igneous systems. During the course of my scientific activity, I explored different aspects of magma mixing using different tools borrowed from the fields of mathematics and physics, together with conventional petrological and volcanological tools. Unfortunately, the COVID-19 emergency almost stopped our research activity and I had the opportunity to have time to look back and realize that this might be the right time to put together the work we had done so far in a more organized way. In doing this I realized two things. The first is that we did a huge amount of work in order to understand the multifaceted nature of magma mixing combining field work, numerical models, and high-temperature experiments. The second is that we are still far from having a coherent conceptual framework accounting for the complexity of this natural process. Nevertheless, I feel that petrologists and volcanologists might appreciate leafing through this little book. I wish to warn the reader that, although this book reports several natural case studies, numerical models and results from experiments, it will not go too much through the mathematical details. The interested readers will find their way to deepen these technical aspects in other sources, including the cited references. On the contrary, this book is mostly intended to stimulate new ideas in students, young and possibly more experienced researches to move further steps to understand what I believe is one of the most important petrological and volcanological processes: the mixing of magmas.

Colle Umberto I, Perugia, Italy
May 2021

Diego Perugini

Acknowledgements

I wanted to be a paleontologist and I ended up an igneous petrologist. The responsible for this is Giampiero Poli, my master and Ph.D. supervisor, and most of all a good friend. I still remember the day I stepped into his office asking about a master project on granitoid rocks from Northern Greece. Needless to say, my interest was mostly on "retsina" than on granites at that time. A few weeks later it was time to fly to Thessaloniki and I went back to Giampiero's office to have some clarifications about fieldwork. He drew a mysterious black blob on a recycled sheet of paper and said: "you have to collect these things in the granitoid masses of the Sithonia Plutonic Complex". I must confess that when I went to Giampiero's office I was a little bit confused. After I left, I was still confused, but on another level. A few days later, I was in the field with George Christofides and George Eleftheriadis and I realized that those black blobs were mafic microgranular enclaves. This was my very first encounter with the extraordinary world of magma mixing. Thank you Giampiero, I would have been a terrible paleontologist and, most of all, I would have missed about twenty years of fabulous research activity we carried out together. This was the time when I learned to do Science.

Roughly in the same period of time another scientist influenced my scientific life: Angelo Peccerillo. The unforgettable discussions with Angelo about geochemical modeling and geodynamics had for me a priceless value.

Around 2000, while I was desperately trying to finish my Ph.D. work, a timid guy knocked to the door of the "Ph.D. student cave". His name is Maurizio Petrelli and he asked for information about a master project on building stones. Lost in my thoughts, I handed him a book on artificial intelligence, asking him to start studying the subject. A week later he showed up with a bunch of working programming scripts. I still have the pleasure to work with Maurizio. Thank you for being what you are: a pragmatic scientist and a friend.

2004 represented a tipping point in my scientific career. This was the year when Jörn Kruhl organized the Fourth International Conference on Fractals and Dynamic Systems in Geoscience. Here I had the opportunity to meet Benoit Mandelbrot and Don Turcotte. Sometimes even a few words pronounced by an eminent scientist can change your mind fueling your interests toward amazing directions. Thank you Jörn, Benoit and Don!

It was during this conference that I also met Kai-Uwe Hess. He invited me to visit the laboratories of Don Dingwell in Munich. I did so and everything changed. Here, I had the privilege to meet Don, Werner Ertel-Ingrisch,

Cristina de Campos and Ulrich Küppers (Ulli). We started a wonderful collaboration that culminated with my stay in Munich for one year as a Humboldt Fellow. In this time, we built the first prototype of chaotic magma mixing device. I still remember our excitement while running the very first experiment. Thank you guys, our scientific meetings at the Oktoberfest were crucial!

A few years later I had the honor to receive a Consolidator Grant from European Research Council Executive Agency. This allowed me to build a new high-temperature lab in Perugia and to organize a wonderful research group. I was lucky enough to recruit excellent researchers, Daniele Morgavi, Maurizio Petrelli and Francesco Vetere, who I wish to thank deeply for their enthusiasm and skills they offered to this adventure.

I am deeply indebted with the Ph.D. students that I had the pleasure to supervise in these years and that shared their time with me in the lab and fieldwork. Among them: Diego González-García, Stefano Rossi, Rebecca Astbury, Kathrin Leager, Joali Paredes Mariño, Alessandro Pisello and Matteo Bisolfati.

I wish also to express all my gratitude to Sandro Conticelli. Although I know he would say "I didn't do anything", we both know that he rescued my scientific life at least twice. Thank you, Sandro.

I also thank all the colleagues who in these years were kind enough to listen to my discussions about magma mixing, chaos and fractals and for making me aware, as vice-director of Department of Physics and Geology of the University of Perugia, of the limits of my patience.

Last, but definitely not least, I wish to thank my wife Laura for giving me the opportunity to refuge in our chicken-coop, which we transformed into a little office for smart-working during the COVID-19 pandemic. She inspired this book and, although she will probably never read it because she is a geomorphologist, her role was decisive.

Contents

Part I

General Overview

What is Magma Mixing?

1

No Geologist worth anything is permanently bound to a desk or laboratory, but the charming notion that true science can only be based on unbiased observation of nature in the raw is mythology. Creative work, in geology and anywhere else, is interaction and synthesis: half-baked ideas from a bar room, rocks in the field, chains of thought from lonely walks, numbers squeezed from rocks in a laboratory, numbers from a calculator riveted to a desk, fancy equipment usually malfunctioning on expensive ships, cheap equipment in the human cranium, arguments before a road cut.

Stephen. J. Gould, "An Urchin in the Storm"

Abstract

This introductory chapter provides a broad overview of magma mixing processes in the volcanic and plutonic environments. The structural, textural and geochemical evidence for magma mixing is briefly discussed, with the aim to provide a general picture about this fundamental, yet relatively poorly known, natural phenomenon. The arguments presented in this chapter will be discussed in greater detail in the next chapters.

1.1 Historical Perspective

Scientific hypotheses generally cross through different stages. The first one is the observation of a new occurrence which is regarded significant, as it differs or contradicts the conventional view about a specific natural phenomenon. The second stage is based on measurements and modelling to quantify, reproduce, and understand the new phenomenon. We can say, therefore, that the infancy of a hypothesis is mostly characterized by observation, whereas its maturity is based on quantification. Infancy and maturity are connected by an intermediate stage in which we realize that observation is not enough to understand satisfactorily the phenomenon and, thus, we need to move to more evolved stages.

In 1851, the chemist Bunsen (1851) published a work suggesting that mixing of magmas might originate most of the compositional variability observed in igneous rocks. Several petrologists criticized ferociously this work and the magma mixing idea was rejected for almost one century (see e.g. Wilcox 1999 for a historical overview). The rise of fractional crystallization as the new paradigm for magma differentiation (Bowen 1915) contributed to bury the magma mixing idea. Nevertheless, around the 1950s the magma mixing idea reinvigorated as the result of the unequivocal indications documented in the rocks (e.g. Bailey and McCallien 1957; Wager and Bailey 1953). After an infancy of about twenty five years, in which a number of observations accumulated (e.g.

D. Perugini, *The Mixing of Magmas*, Advances in Volcanology,
https://doi.org/10.1007/978-3-030-81811-1_1

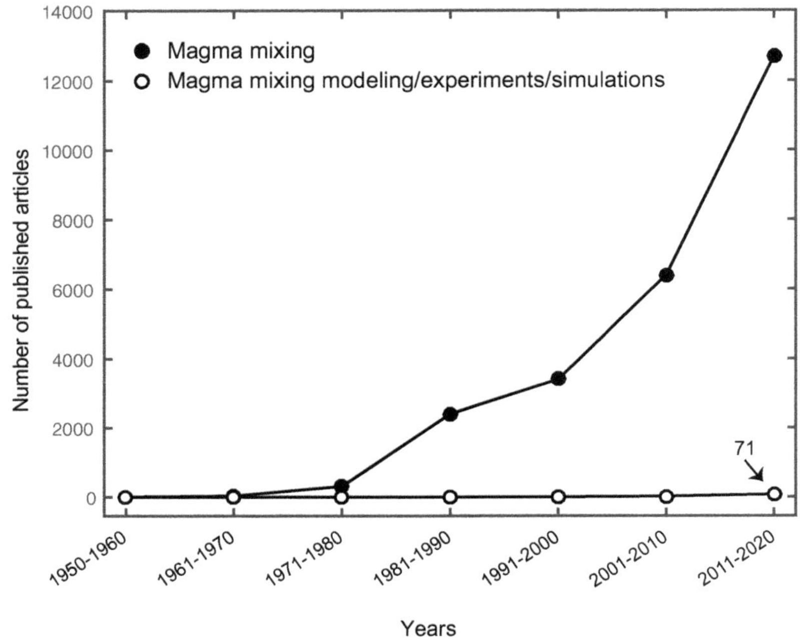

Fig. 1.1 Plot showing the number of papers published on magma mixing from 1950 to 2020. For each ten-year period, the expressions "magma mixing", "magma mixing modelling/experiments/simulations" were used to query the database. *Source data* Google Scholar, January 2021

Walker and Skelhorn 1966; Yoder 1973), this process moved to an intermediate stage (e.g. Huppert et al. 1980; Kouchi and Sunagawa 1985; Oldenburg et al. 1989; Sparks and Marshall 1986; Turner and Campbell 1986; Vernon et al. 1988) and progressively evolved towards maturity (e.g. Bergantz 2000; Jellinek and Kerr 1999; Petrelli et al. 2011, 2018). Figure 1.1, shows the number of articles published on the subject in the time interval 1950–2020. For each ten-year period, the sentences "magma mixing", "magma mixing modelling/experiments/simulations" were used to query the Google Scholar database. The plot displays that starting from 1980s the number of works containing "magma mixing" increases exponentially up to a value of approximately 13000 in 2010–2020. Contrarily, the number of articles containing "modelling, experiments or simulations" remains close to zero up to the beginning of 1990s and, then, it grows very slowly, indicating that the maturity stage may still take time to be reached.

Many works suggest that the magma mixing is a key process in controlling the compositional variation in igneous rocks (e.g. Blundy and Sparks 1992; De Campos et al. 2004; Wiebe 1994) and triggering volcanic explosions (e.g. Leonard et al. 2002; Murphy et al. 1998). However, despite the recognized importance of this process in both igneous petrology and volcanology, and the progresses in experimental and numerical modelling strategies (e.g. Perchuk 1993; De Campos et al. 2011; Laumonier et al. 2014; Perugini et al. 2003; Zimanowski et al. 2004), we are still far from a satisfactory understanding of the physical-chemical mechanisms of magma mixing.

1.2 What Is Magma Mixing?

It has become common practice to apply the term "magma mingling" to refer to the process of physical dispersion (no chemical exchanges are involved) of one or more magmas in another magma. The term "magma mixing" is instead used to indicate that mingling is also accompanied by chemical exchanges (e.g. Flinders and Clemens 1996). Regrettably, such a jargon is rarely used in the literature generating misunderstandings. Although it is often not easy to distinguish between the two processes, we support the idea that pure mingling is not a common process in Nature. As

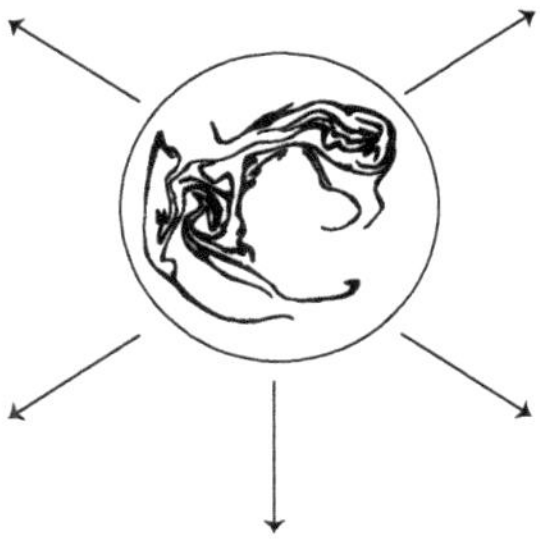

Fig. 1.2 Synopsis of the processes that can potentially generate compositional gradients in igneous systems and, thus, trigger magma mixing

mingling proceeds, the interaction between magmas increases and this increases the probability for chemical exchanges to operate. As we are mostly interested in discussing those circumstances in which the two magmas experience mutual chemical exchanges, in the rest of this book we will use the word "mixing".

It has been commonly accepted that magmas can interact and mix efficiently exclusively when they have a similar rheological behaviour (e.g. Bateman 1995; Poli et al. 1996; Sparks and Marshall 1986). Such physical conditions can occur when: (a) magmas have a similar rheology from the beginning of the interaction process, and (b) different magmas achieve a similar rheology in response to evolutionary processes (e.g. Poli et al. 1996; Sparks and Marshall 1986). However, as we will show in the next chapters, high-temperature experiments clearly indicate that this does not need to be always the case and these ideas must not be taken as dogmas.

The mixing of magmas can occur at any moment during the life cycle of a magmatic system (Fig. 1.2), whenever chemical gradients are formed. This means that mixing can co-occur in association with other petrological processes such as fractional crystallization, assimilation or partial melting, which produce both thermal and chemical gradients. As for fractional crystallization, it is likely to start along the external and cooler walls of magma reservoirs or conduits. This will deplete the melt in those elements that are preferentially incorporated into the crystals generating compositional gradients between the internal and the external parts of the magmatic mass. A further scenario is the process of assimilation of country rocks; here, volumes of magma close to the walls may attain a different chemical composition relative to the most internal portions, again generating compositional gradients that can trigger mixing processes. Also in the case of partial melting compositional gradients might be also produced because of a temperature gradient in a partially molten zone. Melts formed at small melt fractions are more enriched, for example, in Rare Earth Elements (REE) and other incompatible elements relative to those formed at larger melt fractions. These compositional gradients are the prerequisite for triggering mixing processes. Besides, whenever coeval melts with different compositions are formed, it is very likely that they will be transported to different crustal levels. The different melts can crosscut their mutual trajectories and, therefore, mix. Noteworthy, all these processes can act simultaneously, amplifying the effect of mixing and triggering a considerable inhomogeneity in magmatic masses.

Therefore, although the most striking evidence of magma mixing in the rocks is the occurrence of at least two different magmas, commonly with

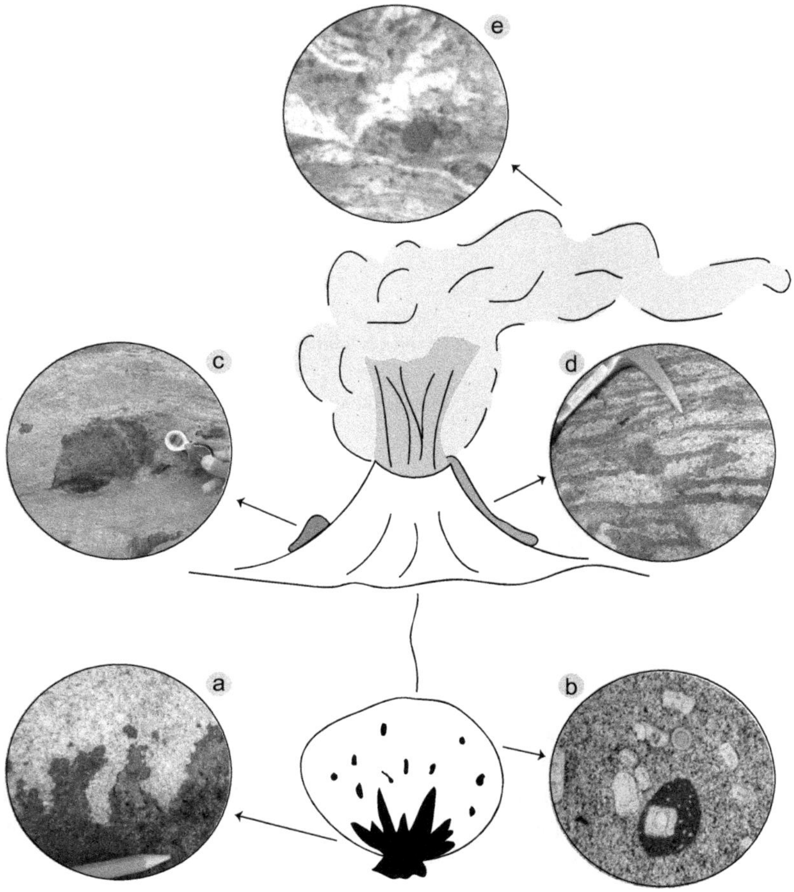

Fig. 1.3 Occurrence of magma mixing in different igneous environments. Shown pictures are from: **a** Terra Nova Intrusive Complex (Northern Victoria Land, Antarctica); **b** Monte Capanne pluton (Island of Elba, Tuscan Magmatic Province, Italy); **c** Monte Guardia lava domes (Lipari, Aeolian Islands, Italy); **d** a lava flow (Island of Lesvos, Greece); **e** Pollara pyroclastic deposits (Salina, Aeolian Islands, Italy)

different compositions, the concept of magma mixing can be extended to consider all those cases where chemical gradients are present in magma bodies.

1.3 The Witnesses of Magma Mixing in the Rocks

Magma mixing can take place in any geological configuration, from buried magma chambers to volcanic plumbing systems (e.g. Fig. 1.3), and generate a large diversity of patterns in the rocks. These can be classified into three main groups:

1. flow structures;
2. magmatic enclaves;
3. physical-chemical disequilibrium in minerals.

While group 1 and 2 are typical for magma mixing, group 3 structures do not document univocally this process. In fact, especially in volcanic rocks, physical and chemical disequilibrium in minerals can be also caused by other processes such as decompression during magma ascent toward the Earth's surface (e.g. Stewart and Fowler 2001). Therefore, caution must be used to infer magma mixing processes based only on the study of compositional variations in minerals.

1.3.1 Flow Structures

Flow structures indicate portions of magma having spatial continuity dispersed within another magma. These can be recognized in the rocks as alternations of bands with different colors.

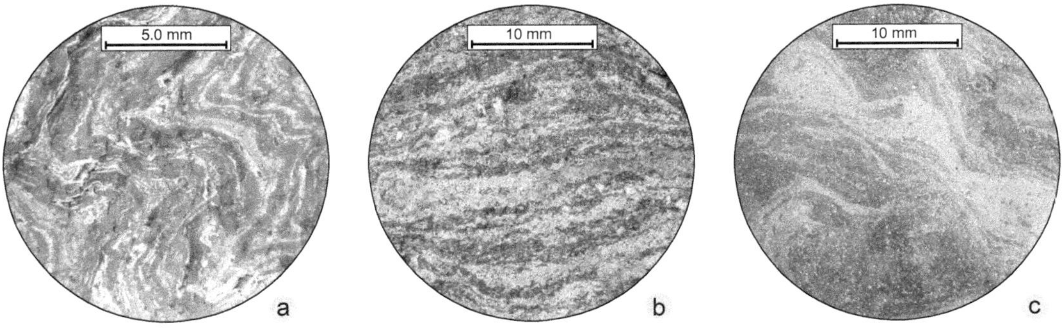

Fig. 1.4 Examples of flow structures occurring in volcanic rocks; **a** and **b** flow structures from a lava flow from the Island of Lesvos; **c** structures in a lava flow on the Island of Vulcano (Aeolian islands, Italy)

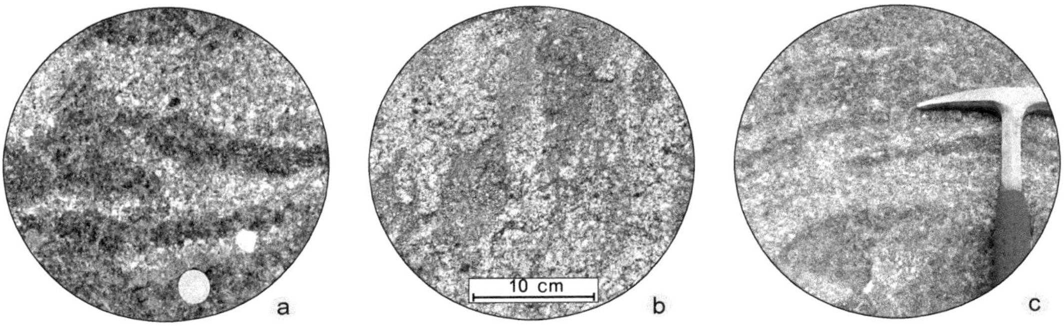

Fig. 1.5 Examples of flow structures occurring in plutonic rocks; **a** Sithonia Plutonic Complex (Greece); **b** and **c** the pluton of Philippi (Greece)

Flow structures mark the evolution of flow fields during the mixing process and produce rocks where the mixing dynamics are fossilized. Figure 1.4 shows some examples of flow structures in volcanic rocks. In particular, Fig. 1.4a and b show flow structures in lava flows from the Island of Lesvos (North-Eastern Greece; Perugini et al. 2003); Fig. 1.4c displays structures in a lava flow from the Island of Vulcano (Aeolian islands, Southern Italy; Perugini et al. 2003).

Flow structures are generally less abundant in plutonic rocks (e.g. Didier and Barbarin 1991); Fig. 1.5 shows some of these structures in the plutons of the Sithonia Plutonic Complex (Greece; e.g. Perugini et al. 2003) and Philippi (Greece; e.g. Eleftheriadis et al. 1995).

1.3.2 Magmatic Enclaves

The term magmatic enclave is used to identify a portion of magma embedded within another magma (e.g. Bacon 1986; Didier and Barbarin 1991). Although contacts between enclaves and host rocks can be sharp, the margins of many enclaves are often crenulate. Some enclaves in volcanic rocks are shown in Fig. 1.6. Figure 1.6a shows enclaves in lava flow on the Island of Salina (Aeolian Islands; e.g. De Rosa et al. 1996). Figure 1.6b reports magmatic enclaves from Monte Guardia lava domes (Lipari, Aeolian Islands). Figure 1.6c shows portion of an enclave from the Grizzly Lake lava flow (Yellowstone National Park, USA).

In the plutonic environment magmatic enclaves are considered as one of the most important evidence for magma mixing. These enclaves are generally named with the acronym MME (Mafic Microgranular Enclaves; e.g. Didier and Barbarin 1991). MME have typically a smaller grain size and less evolved geochemical and mineralogical characteristics relative to the host rock. Figure 1.7a and b show MMEs occurring in the

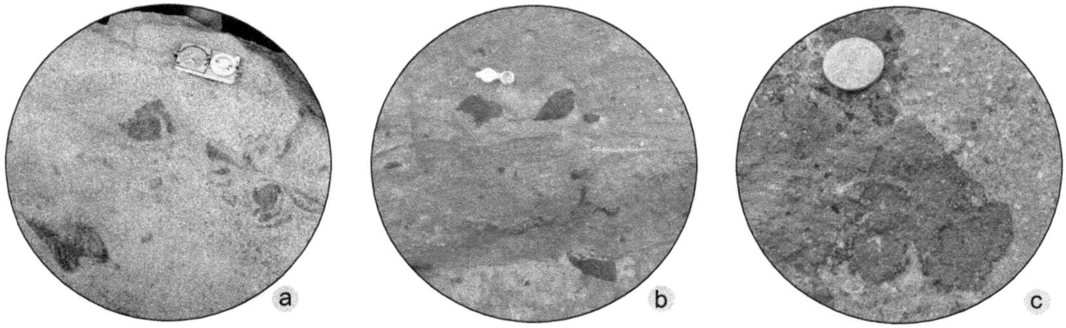

Fig. 1.6 Examples of enclaves in volcanic rocks; **a** enclaves from the Porri lava flows on the Island of Salina (Aeolian Islands, Italy); **b** enclaves from the Monte Guardia lava domes (Island of Lipari, Aeolian Islands, Italy); **c** enclave from Grizzly Lake lava flow (Yellowstone National Park, USA)

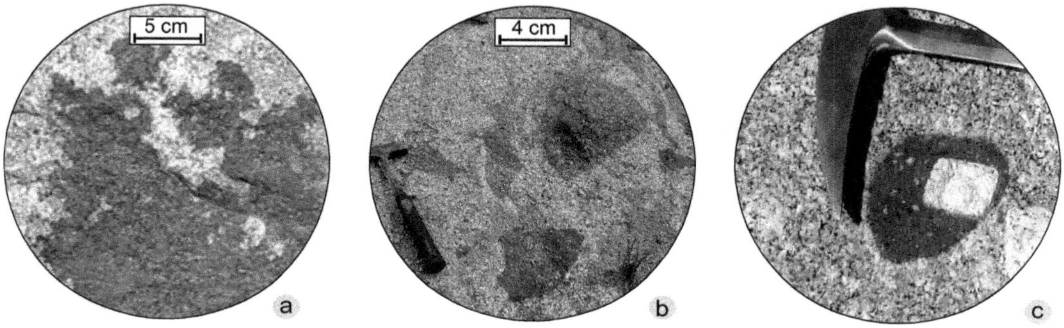

Fig. 1.7 Examples of Mafic Microgranular Enclaves (MME) in plutonic rocks; **a** pluton of Aztech Wash (Nevada, USA) and **b** pluton of Philippi (Greece); **c** pluton of Monte Capanne (Island of Elba, Tuscan Magmatic Province, Italy), also showing the occurrence of crystals belonging to the host magma in the MME

Aztech Wash pluton (Nevada, USA) and the granitoid rocks of Philippi (Greece; Perugini et al. 2003), respectively. Figure 1.7c displays an enclave in the pluton of Monte Capanne (Tuscan Magmatic Province, Elba Island, Italy; e.g. Dini et al. 2002). Pictures display that, despite contacts between host rocks and enclaves may appear quite sharp, enclave boundaries are often interrupted by inter-digitations with the host magma. Figure 1.7c also reports the presence of minerals (K-feldspar) of the host magma in the MME. The occurrence of mineral phases belonging to the host magma showing chemical-physical disequilibrium is a common feature of MME.

In some granitoid masses, especially those considered to represent the zone of injection of the mafic magma into the felsic one, enclaves can show a wide variety of morphology of the contact interfaces (e.g. Perugini et al. 2005; Perugini and Poli 2005; Wiebe et al. 2001). Some examples occurring in the Terra Nova Intrusive Complex (Northern Victoria Land, Antarctica; Perugini and Poli 2005) and the Aztec Wash Pluton (Nevada, USA; Perugini and Poli 2011) are shown in Fig. 1.8. In some cases these morphologies can be also observed in volcanic rocks, although their interpretation requires caution (e.g. Perugini et al. 2007).

1.3.3 Chemical-Physical Disequilibrium in Crystals

There are virtually endless possibilities for crystals to show physical-chemical disequilibrium and a systematic review can be found in Hibbard (1994). Here we will focus on resorption and zoning textures (e.g. Anderson and Anderson

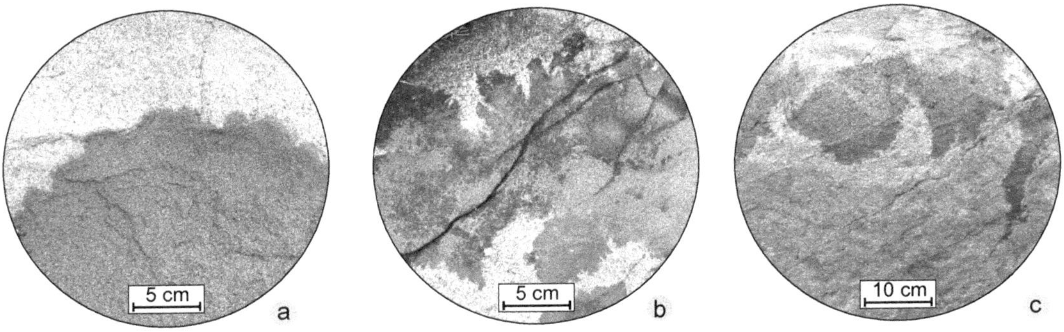

Fig. 1.8 MME in the Terra Nova Intrusive Complex (**a** and **b**) and the Aztec Wash Pluton (**c**). The pictures display the large variability of contact morphology between the MME and the host rock

1984; Ginibre et al. 2002; Wallace and Bergantz 2002). Figure 1.9a shows a representative plagioclase crystal from the Sithonia Plutonic Complex (Northern Greece; e.g. Perugini et al. 2003). Figure 1.9b and c display disequilibrium textures for a plagioclase and a diopside crystal from Capraia (Tuscan Magmatic Province, Italy; Perugini et al. 2005) volcanites and the Santa Venera lava flow (Mt. Etna, Italy; Perugini et al. 2003), respectively.

The invasion of a felsic magma chamber by a mafic magma produces dramatic thermal and compositional disequilibrium (e.g. Bateman 1995; Sparks and Marshall 1986). Minerals already present in the magmas react to this disequilibrium. For example, minerals in the low-temperature felsic magma will likely experience remelting/resorption because of the increase in temperature. Nevertheless, depending on the relative timing of resorption kinetics and attainment of new thermal equilibrium, the resorption process may not be complete. Resorbed grains can act as nuclei for the growth of the same mineral species, but with a different composition (e.g. Fig. 1.9). The compositional heterogeneity triggered by the mixing process can also play a key role in determining the degree of resorption and growth of crystals (e.g. Anderson and Anderson 1984; Perugini et al. 2005). These processes can reiterate in time leading to the production of complex oscillatory zoning patterns. Their association with mafic enclaves and/or flow structures makes the study of mineral disequilibrium an extraordinary tool to investigate the short length-scale propagation of the compositional heterogeneity in the igneous system (e.g. Ginibre et al. 2002; Perugini et al. 2005).

1.3.4 Geochemical Evidence for Magma Mixing

The above discussed textural and structural evidence of magma mixing is the consequence of variations in the geochemical composition of the magmatic system. There is general consensus that linear trends in binary plots (for any couple of major or trace elements) point to magma mixing processes (e.g. Fourcade and Allegre 1981; Rollinson 1993). Based on this assumption the geochemical modeling is thus commonly performed using linear equations. In the case of interaction between mafic (high-temperature) and felsic (low-temperature) magmas, these linear model are not adequate to account for further processes that may occur in the magmatic system. These include crystallization of the mafic magma as well as remelting or dissolution of crystals that were already present in either magmas. To overcome these difficulties, some authors (e.g. Poli and Tommasini 1999; Poli et al. 1996) proposed more sophisticated models such as the so-called "Mixing plus Fractional Crystallization" (MFC) model. This model has the same mathematical formulation of the "Assimilation and Fractional Crystallization" AFC model (DePaolo 1981). In this model the assimilant is represented by a felsic magma in

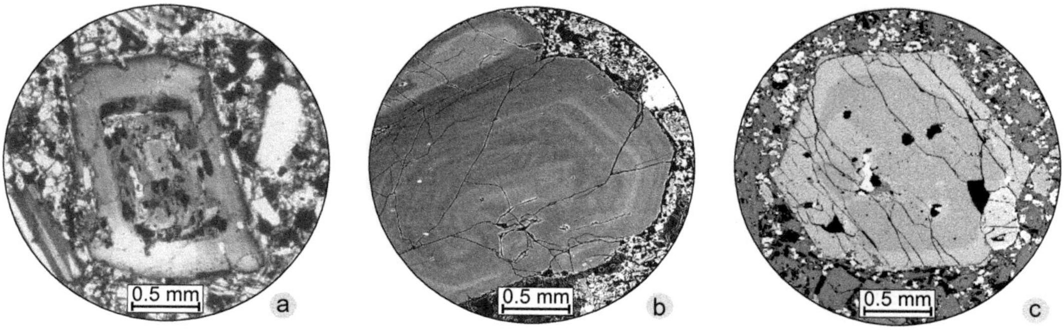

Fig. 1.9 Representative samples of minerals showing resorption and zoning due to magma mixing; **a** plagioclase crystal (optical microscope image) in the rocks of the Sithonia Plutonic Complex (Northern Greece) **b** BSE (Back-Scattered Electron) picture of plagioclase crystal with oscillatory zoning from the Island of Capraia (Tuscan Magmatic Province, Italy); **c** zoned diopside crystal (optical microscope image) from the Santa Venera lava flow (Mt. Etna, Italy)

which a mafic magma is dispersed. During mixing the mafic magma experiences fractional crystallization resulting in a progressively more evolved melt.

The major problem with both of the above geochemical models is that they are expected to generate monotonously increasing (or decreasing) trends in inter-elemental plots. In addition, these models work under the assumption that the magmatic system is completely homogeneous everywhere and at any time during its evolution. As will be shown in the following chapters, these assumptions can generate oversimplifications with two main disadvantages: (i) the system is modelled with a large degree of approximation and (ii) crucial information to understand the evolution in time and space of the magmatic system can be lost.

1.4 Concluding Remarks

Magma mixing can occur at any stage of life of a magmatic system, anytime compositional gradients are formed. This means that any igneous mass can be potentially affected by this process.

We will see in the next chapters that the use of conceptual models based on Chaos Theory and Fractal Geometry will allow us to define a new framework to face the challenging task of understanding this complex, yet fundamental, petrological and volcanological process.

Application of these methods, in combination with conventional petrological techniques, will help us to understand several structural, textural, and geochemical features that are still poorly understood.

A possible approach to reach this target is to establish a link between the evidence in the rocks and fluid dynamics, using numerical models and high-temperature experiments. If on one hand this will allow us to move further steps in understanding the mixing of magmas, on the other hand it will open a variety of new intriguing questions. Among them, probably the most important one is: will we ever have the opportunity to understand exhaustively the complex evolution of igneous systems, avoiding the use of oversimplified conceptual models? Responding to this question is undoubtedly not easy. However, as we will try to show in this book, the investigation of magma mixing cannot be separated from the study of the dynamics controlling the evolution of igneous systems in space and time.

In the next two chapters we will briefly introduce some basic concepts of Chaos Theory and Fractal Geometry, as well as considerations about chaotic mixing of fluids. The goal is to provide to the reader the basis for reading and following at best the arguments presented in Part II, III and IV.

References

Anderson AY, Anderson AT Jr (1984) Probable relations between plagioclase zoning and magma dynamics, Fuego Volcano, Guatemala. Amer Mineralog 69:660–676

Bacon CR (1986) Magmatic inclusions in silicic and intermediate volcanic rocks. J Geophys Res 91.B6:6091. ISSN: 0148-0227. https://doi.org/10.1029/jb091ib06p06091

Bailey EB, McCallien WJ (1957) Composite minor intrusions, and the slieve gullion complex, ireland. Geolog J 1(6):466–501. ISSN: 10991034. https://doi.org/10.1002/gj.3350010602

Bateman R (1995) The interplay between crystallization, replenishment and hybridization in large felsic magma chambers (English). Earth-Sci Rev 39(1–2):91–106. ISSN: 00128252. https://doi.org/10.1016/0012-8252(95)00003-S

Bergantz GW (2000) On the dynamics of magma mixing by reintrusion: implications for pluton assembly processes (English). J Struct Geol 22(9):1297–1309. ISSN: 01918141. https://doi.org/10.1016/S0191-8141(00)00053-5

Blundy JD, Sparks RSJ (1992) Petrogenesis of mafic inclusions in granitoids of the Adamello Massif, Italy. J Petrol 33(5):1039–1104. https://doi.org/10.1093/petrology/33.5.1039

Bowen NL (1915) The later stages of the evolution of the igneous rocks. J Geol 23(S8):1–91. ISSN: 0022-1376. https://doi.org/10.1086/622298

Bunsen R (1851) Ueber die Processe der vulkanischen Gesteinsbildungen Islands. Annalen der Physik 159(6):197–272. ISSN: 15213889. https://doi.org/10.1002/andp.18511590602

De Campos CPP et al (2011) Enhancement of magma mixing efficiency by chaotic dynamics: an experimental study (English). Contributions to Mineral Petrol 161(6):863–881. https://doi.org/10.1007/s00410-010-0569-0

De Campos CP, et al (2004) Decoupled convection cells from mixing experiments with alkaline melts from Phlegrean Fields. Chem Geol 213(1–3):227–251. ISSN: 00092541. https://doi.org/10.1016/j.chemgeo.2004.08.045

De Rosa R, Mazzuoli R, Ventura G (1996) Relationships between deformation and mixing processes in lava flows: a case study from Salina (Aeolian Islands, Tyrrhenian Sea). Bull Volcanol 58:286–297

DePaolo DJ (1981) Trace element and isotopic effects of combined wallrock assimilation and fractional crystallization. Earth Planet Sci Lett 53(2):189–202. ISSN: 0012821X. https://doi.org/10.1016/0012-821X(81)90153-9

Didier J, Barbarin B (1991) Enclaves and granite petrology - Developments in petrology, vol 13. Elsevier, Amsterdam

Dini A, et al (2002) The magmatic evolution of the late Miocene laccolithpluton-dyke granitic complex of Elba Island, Italy (English). Geolog Mag 139(3):257–279. ISSN: 00167568. https://doi.org/10.1017/S0016756802006556

Eleftheriadis G, Pe-Piper G, Christofides G (1995) Petrology of the Philippi granitoid rocks and their microgranular enclaves (East Macedonia, North Greece). Geolog Soc Greece 4:512–517

Flinders J, Clemens JD (1996) Non-linear dynamics, chaos, complexity and enclaves in granitoid magmas (English). Trans R Soc Edinburgh-Earth Sci 87:217–223

Fourcade S, Allegre CJ (1981) Trace element behaviour in granite genesis: a case study the calc-alkaline plutonic association from the Querigut Complex (Pyrenees, France). Contrib Mineral Petrol 76:177–195

Ginibre C, Kronz A, Worner G (2002) High-resolution quantitative imaging of plagioclase composition using accumulated backscattered electron images: new constraints on oscillatory zoning (English). Contr Mineral Petrol 142(4):436–448

Hibbard MJ (1994) Petrography to petrogenesis. Prentice Hall, Hoboken

Huppert HE, Stephen R, Sparks J (1980) The fluid dynamics of a basaltic magma chamber replenished by influx of hot, dense ultrabasic magma. Technical report 1980, p 289

Jellinek MA, Kerr RC (1999) Mixing and compositional layering produced by natural convection. Part 2. Applications to the differentiation of basaltic and silicic magma chambers, and komatiite lava flows. J Geophys Res 104:7203–7218

Kouchi A, Sunagawa I (1985) A model for mixing basaltic and dacitic magmas as deduced from experimental-data (English). Contrib Mineral Petrol 89(1):17–23

Laumonier M, et al (2014) Experimental simulation of magma mixing at high pressure. Lithos 196–197:281–300. ISSN: 00244937. https://doi.org/10.1016/j.lithos.2014.02.016

Leonard GS, et al (2002) Basalt triggering of the c. AD 1305 Kaharoa rhyolite eruption, Tarawera Volcanic Complex, New Zealand". English. J Volcanol Geother Res 115(3–4):461–486. ISSN: 03770273. https://doi.org/10.1016/S0377-0273(01)00326-2

Murphy MD, et al (1998) The role of magma mixing in triggering the current eruption at the Soufriere Hills Volcano, Montserrat, West Indies. Geophys Res Lett 25(18):3433–3436. ISSN: 00948276. https://doi.org/10.1029/98GL00713

Oldenburg CM, et al (1989) Dynamic mixing in magma bodies - theory, simulations, and implications (English). J Geophys Res-Solid Earth Planets 94(B7):9215–9236

Perchuk LL (1993) Experimental studies of magma mixing. Int Geol Rev 35(8):721–738. ISSN: 19382839. https://doi.org/10.1080/00206819309465553

Perugini D, et al (2003) Magma mixing in the Sithonia Plutonic complex, Greece: evidence from mafic microgranular enclaves (English). Mineral Petrol 78(3–4):173–200. ISSN: 09300708. https://doi.org/10.1007/s00710-002-0225-0

Perugini D, et al (2003) The role of chaotic dynamics and flow fields in the development of disequilibrium textures in volcanic rocks (English). J Petrol 44(4):733–756

Perugini D, Poli G (2005) Viscous fingering during replenishment of felsic magma chambers by continuous inputs of mafic magmas: field evidence and fluid-mechanics experiments. Geology 33(1):5–8

Perugini D, Poli G (2011) Frontiers: intrusion of mafic magmas into felsic magma chambers: New insights from natural outcrops and fluidmechanics experiments. Ital J Geosci 130(1):3–15

Perugini D, Poli G, Mazzuoli R (2003) Chaotic advection, fractals and diffusion during mixing of magmas: evidence from lava flows. J Volcanol Geother Res 124(3–4):255–279. ISSN: 03770273. https://doi.org/10.1016/S0377-0273(03)00098-2

Perugini D, Poli G, Rocchi S (2005) Development of viscous fingering between mafic and felsic magmas: evidence from the Terra Nova Intrusive Complex (Antarctica) (English). Mineral Petrol 83(3–4):151–166. ISSN: 09300708. https://doi.org/10.1007/s00710-004-0064-2

Perugini D, Poli G, Valentini L (2005) Strange attractors in plagioclase oscillatory zoning: petrological implications (English). Contrib Mineral Petrol 149(4):482–497

Perugini D, Valentini L, Poli G (2007) Insights into magma chamber processes from the analysis of size distribution of enclaves in lava flows: a case study from Vulcano Island (Southern Italy). J Volcanol Geother Res 166(3–4):193–203. ISSN: 03770273. https://doi.org/10.1016/j.jvolgeores.2007.07.017

Petrelli M, et al (2018) Timescales of water accumulation in magmas and implications for short warning times of explosive eruptions. Nat Commun 9(1):1–14. ISSN: 20411723. https://doi.org/10.1038/s41467-018-02987-6

Petrelli M, Perugini D, Poli G (2011) Transition to chaos and implications for time-scales of magma hybridization during mixing processes in magma chambers. Lithos 125(1–2):211–220. ISSN: 00244937. https://doi.org/10.1016/j.lithos.2011.02.007

Poli G, Tommasini S (1999) Geochemical modeling of acidbasic magma interaction in the Sardinia-Corsica Batholith: the case study of Sarrabus, southeastern Sardinia, Italy. Lithos 46(3):553–571. ISSN: 00244937. https://doi.org/10.1016/S0024-4937(98)00082-6

Poli G, Tommasini S, Halliday AN (1996) Trace element and isotopic exchange during acid-basic magma interaction processes (English). Trans R Soc Edinburgh-Earth Sci 87(1–2):225–232. ISSN: 02635933. https://doi.org/10.1017/s0263593300006635

Rollinson H (1993) Using geochemical data: evaluation, presentation, interpretation. Longman Scientific and Technical, London, p 352

Sparks RSJ, Marshall LA (1986) Thermal and mechanical constraints on mixing between mafic and silicic magmas. J Volcanol Geother Res 29(1–4):99–124. ISSN: 03770273. https://doi.org/10.1016/0377-0273(86)90041-7

Stewart ML, Fowler AD (2001) The nature occurence of discrete zoning in plagioclase from recently erupted andesitic volcanic rocks, Montserrat (English). J Volcanol Geother Res 106(3–4):243–253. https://doi.org/10.1016/S0377-0273(00)00240-7

Turner JS, Campbell IH (1986) Convection and mixing in magma chambers. Earth Sci Rev 23(4):255–352. ISSN: 00128252. https://doi.org/10.1016/0012-8252(86)90015-2

Vernon RH, Etheridge MA, Wall VJ (1988) Shape and microstructure of microgranitoid enclaves: Indicators of magma mingling and flow. Lithos 22(1):1–11. ISSN: 00244937. https://doi.org/10.1016/0024-4937(88)90024-2

Wager LR, Bailey EB (1953) Basic magma chilled against acid magma. Nature 172(4367):68–69. ISSN: 00280836. https://doi.org/10.1038/172068a0

Walker GPL, Skelhorn RR (1966) Some associations of acid and basic igneous rocks. Earth Sci Rev 2(C):93–109. ISSN: 00128252. https://doi.org/10.1016/0012-8252(66)90024-9

Wallace GS, Bergantz GW (2002) Wavelet-based correlation (WBC) of zoned crystal populations and magma mixing (English). Earth Planet Sci Lett 202(1):133–145. ISSN: 0012821X. https://doi.org/10.1016/S0012-821X(02)00762-8

Wiebe RA (1994) Silicic magma chambers as traps for basaltic magmas: the Cadillac Mountain intrusive complex, Mount Desert Island, Maine. J Geol 102(4):423–437. ISSN: 00221376. https://doi.org/10.1086/629684

Wiebe RA, Frey H, Hawkins DP (2001) Basaltic pillow mounds inthe Vinalhaven intrusion, Maine. J Volcanol Geother Res 107(1–3):171–184. ISSN: 03770273. https://doi.org/10.1016/S0377-0273(00)00253-5

Wilcox RE (1999) The idea of magma mixing: history of a struggle for acceptance. J Geol 107(4):421–432. ISSN: 00221376. https://doi.org/10.1086/314357

Yoder H (1973) Contemporaneous basaltic and rhyolitic magmas. Amer Mineral 58:153–171

Zimanowski B, Büttner R, Koopmann A (2004) Experiments on magma mixing. Geophys Res Lett 31(9). ISSN: 00948276. https://doi.org/10.1029/2004GL019687

2 Chaos Theory and Fractal Geometry

You've never heard of Chaos Theory? Non-linear equations? Strange attractors? Ms. Sattler, I refuse to believe you're not familiar with the concept of attraction.

Michael Cricton, "Jurassic Park"

Abstract

This chapter provides an introduction to chaos theory and fractal geometry to aid the reader in following the arguments presented in the next chapters. Fractal geometry and chaos theory represent the building blocks for assembling a holistic conceptual model to study the complexity of magma mixing processes. They are tools, similar to conventional petrological techniques to quantify structures and geochemical patterns that are difficult to study using only classic petrological methods. Chaos theory and fractal geometry equip petrologists and volcanologists with a new pair of glasses to observe a novel world, which contains ideas that may prove fundamental in the challenge of deciphering the impact of magma mixing on our understanding of planetary differentiation and volcanic activity.

2.1 Introduction

The aim of this chapter is to provide an introduction to chaos theory and fractal geometry to aid the reader in following the arguments presented in the next chapters. The text does not pretend to be an exhaustive overview of the immense literature about these subjects. Rather, it introduces the main concepts and the interested reader is invited to refer to the cited references for further details.

Fractal geometry and chaos theory represent the building blocks for assembling a holistic conceptual model to study the complexity of magma mixing processes. They should be considered as tools, similar to conventional petrological techniques, such as geochemical models or sophisticated analytical equipment, to quantify structures and geochemical patterns that are difficult to study using only classic petrological methods. Chaos theory and fractal geometry equip petrologists and volcanologists with a new pair of glasses to observe a novel world, which contains ideas that may prove fundamental in the challenge of deciphering the impact of magma mixing on our understanding of planetary differentiation and volcanic activity.

2.2 Chaos Theory

In the book of Theogony, Hesiod (ca. 700 B.C.) defines Chaos as the "dark and shapeless void

D. Perugini, *The Mixing of Magmas*, Advances in Volcanology,
https://doi.org/10.1007/978-3-030-81811-1_2

made up of non-matter". It was the very first creature created by the Gods. Afterwards, they created a female being called Night. Chaos and Night were supposed to join to create the whole universe. Such a universe was perfectly in line with the concept of an unpredictable and sometimes malignant Nature, subjugated to the whims of the Gods. Around the 5th–6th century B.C. emerged the new idea that the universe may have had an internal order and that Nature obeyed very precise rules dictated by the theories of order and perfection of the classical Greek school. This ordered view of the universe was called Cosmos, which means beauty, order and harmony. Chaos and Cosmos were therefore antagonists, one representing the unknown and the unpredictable, the other the mirror of perfection.

The vision of a Nature obeying simple rules continued over the centuries, exploding in 1687 in the work of Isac Newton "Philisophiae Naturalis Principia Mathematica", where everything that happened in Nature could be explained with the laws of classical mechanics. This scientific paradigm was considered valid until the beginning of the 20th century, when scientists started to realize that small uncertainties about the knowledge of the conditions of a natural system could cause completely unpredictable evolutions in a very short time.

The opinions of P. S. Laplace and E. Poincaré, who lived in different historical moments reflect the change of perspective in science before and after this realization.

P. S. Laplace (1776): "We may regard the present state of the universe as the effect of its past and the cause of its future. An intellect which at a certain moment would know all forces that set nature in motion, and all positions of all items of which nature is composed, would embrace in a single formula the movements of the greatest bodies of the universe and those of the tiniest atom; for such an intellect nothing would be uncertain and the future just like the past would be present before its eyes."

E. Poincaré (1903): "If we knew exactly the laws of nature and the conditions of the universe at the initial moment, we could predict exactly its evolution. But even in the case that the natural laws had no longer any secret for us, we could still only know the initial conditions approximately. It may happen that small differences in the initial conditions produce very great ones in the final phenomena. A small error in the former will produce an enormous error in the latter. Prediction becomes impossible, and we have a fortuitous phenomenon."

Laplace's statement reflects the scientific paradigm of his time: the knowledge of all the conditions of a system at a given instant are a necessary and sufficient condition to predict its behavior with extreme precision. Poincaré's statement contains what in modern science is the concept of chaos. Small uncertainties in the knowledge of the conditions of a natural system can produce very large uncertainties in a very short time and cause an enormous loss of information. This does not mean that Newton's laws are no longer valid. They are still valid, but the uncertainty due to the approximate knowledge of the initial conditions of a system implies that it is very hard to predict with certainty its evolution.

2.3 Dynamical Systems

The wider conceptual framework in which chaos emerges is the so-called "dynamical systems theory". A dynamical system consists of two fundamental parts: the characteristics of its state, that is, the essential information about the system, and the dynamics, a rule that describes its state over time. Its evolution can be visualized in the so-called "phase space". In dynamical system theory, a phase space is a space in which all possible states of a system are represented. For mechanical systems, for example, the phase space usually consists of all possible values of position and velocity. Although this concept may appear too much theoretical for our scopes, it will be clear later in this book that the concept of phase space in fluid mixing is fundamental. Indeed, phase space and physical space in which the mixing process occurs coincide. The phase space is a powerful tool for describing the behavior of dynamical systems because it allows us to represent their evolution in a geometric form (e.g. Alligood et al. 1996).

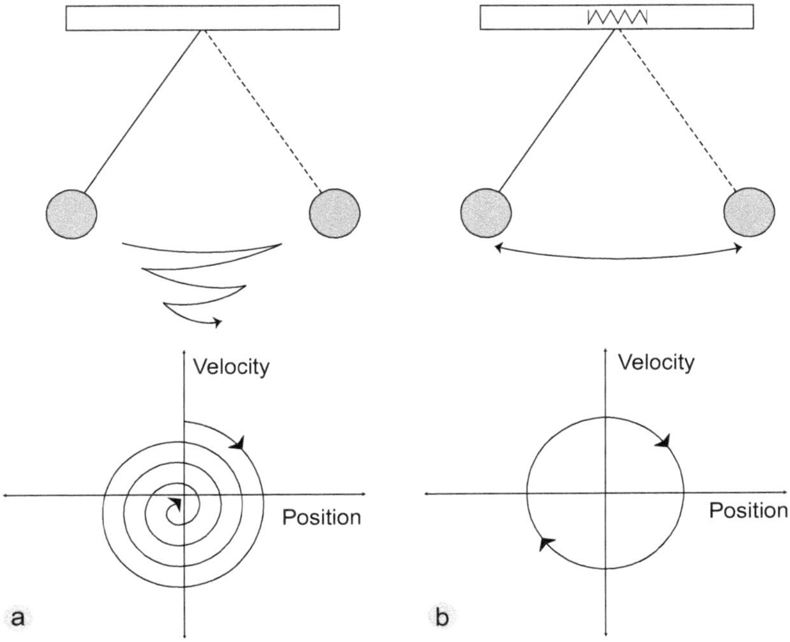

Fig. 2.1 A pendulum subject to friction will stop progressively its motion (**a**); the corresponding attractor is a fixed point. If the energy loss of the pendulum is compensated by a spring, the motion will be constant; the corresponding attractor is a limit cycle (**b**)

For example, a pendulum subject to friction sooner or later will stop its motion; in the phase space this corresponds to an orbit tending towards a point (Fig. 2.1a). The point does not move, it is a "fixed point", and since it attracts nearby orbits, it is called an "attractor" (e.g. Baker and Gollub 1996; Alligood et al. 1996). If we give a weak push to the pendulum, it will return to the same attractor (i.e. the fixed-point). Systems that over time reach a state of rest can be characterized by a fixed point in the state space. This is a particular case of a very general phenomenon that occurs when, for example, friction or viscosity cause the orbits of the dynamical system to be attracted to a region smaller than the entire phase space. These regions are also called attractors. In general terms, an attractor is the region of phase space in which the behavior of a system stabilizes or is attracted to.

Some systems do not tend towards rest, but, on the contrary, pass cyclically through a periodic succession of states. An example is constituted by a pendulum where energy loss due to friction is integrated by a spring (Fig. 2.1b). The pendulum continues to repeat the same motion. In the phase space this motion corresponds to a periodic orbit or cycle (Fig. 2.1b). Regardless of how the pendulum is placed in oscillation, the cycle representing the limit to which it tends in the long term will be always the same. Attractors of this type are therefore called "limit cycles" (e.g. Addison 1997; Baker and Gollub 1996).

The set of points that evolve towards the same attractor constitutes its "basin of attraction". A dynamical system can have multiple attractors. They can exist excluding each other or they can coexist within the same system. In the first case, different initial conditions can drive the system towards different attractors. For example, the pendulum has two basins of attraction. If the movement of the pendulum from the rest position is small, it returns to this position. If, on the other hand, the displacement is large, the pendulum can start swinging in a stable way. In the second case, different attractors can coexist in different regions of the system. This is the case, for example, of systems in which mixing processes between fluids take place. Such systems are of primary interest in this book and will be discussed in detail in the next chapter.

Another common type of attractor is the torus (Fig. 2.2c), which resembles the surface of a donut. This shape can describe a motion

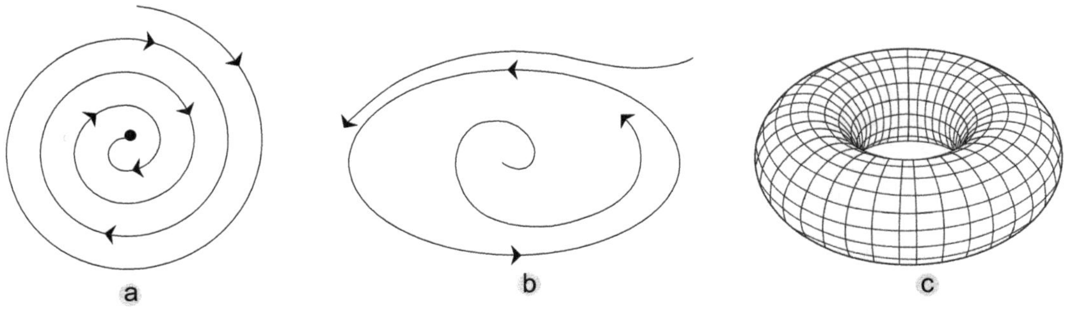

Fig. 2.2 Examples of attractors: **a** fixed-point **b** limit cycle and **c** torus

consisting of two independent oscillations (e.g. Addison 1997; Baker and Gollub 1996). In the phase space the orbits wrap around the torus. One of the two frequencies is determined by the speed with which the orbit moves around the torus in the shorter dimension, while the other depends on the speed with which the orbit moves along the larger dimension. Even though the orbits do not repeat exactly, orbits that start close together remain close together and long-term predictability is ensured.

2.4 Strange Attractors

Until about the 1960s, the only known attractors were fixed points, limit cycles and tori (Fig. 2.2). In 1963 Edward N. Lorenz discovered an example of a system with a few degrees of freedom but with a complex behavior (Lorenz 1963). Lorenz was trying to understand the unpredictability of weather conditions. Starting from the equations of motion of a fluid in convection, after some simplifications he obtained a system with only three degrees of freedom. The system, however, behaved in such a "random" way that it could not be adequately characterized by any of the three attractors known until then. The attractor he observed, and which is now called the Lorenz attractor, was the first example of a chaotic attractor, or "strange attractor" (e.g. Addison 1997; Baker and Gollub 1996).

Figure 2.3 shows sections of the Lorenz attractor along the $x-y$, $x-z$ and $y-z$ planes. These planes represent Poincaré sections of the attractor. A Poincaré section is a plane that passes through the attractor and on which it is possible to visualize some characteristics that, given the complex structure and the large number of dimensions of the attractor, would not be possible to observe (e.g. McCauley 1993). In the case of the Lorenz system, Poincaré sections show that the orbits that describe the attractor draw two "wings" containing a very large number of orbits that never intersect each other.

Lorenz studied the fundamental mechanism originating the observed randomness and discovered that microscopic perturbations of the system were amplified to the point of interfering with the macroscopic behavior. Two orbits corresponding to near initial conditions diverged exponentially and, therefore, remained close each other only for a short time. This behavior is different in non-chaotic attractors, for which the orbits remain close to each other and small errors remain small.

The randomness of the Lorenz dynamical system acquired a technical name: "sensitive dependence on the initial conditions", also popularized as the "butterfly effect". The sensitive dependence on the initial conditions is a fundamental property of chaotic dynamical systems.

2.5 Stretching and Folding: The Fingerprint of Chaos

The key to interpreting chaotic behavior lies in understanding a simple "stretching and folding" operation taking place in the phase space of a dynamical system (Addison 1997; Baker and Gollub

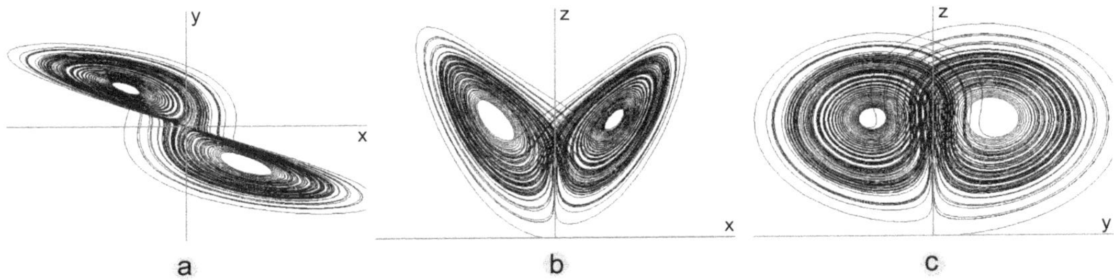

Fig. 2.3 Sections of the Lorenz attractor along the $x-y$, $x-z$ and $y-z$ planes

1996). The exponential divergence observed by Lorenz is a local phenomenon. In fact, since the size of the attractors is finite, two orbits located on a chaotic attractor cannot diverge forever. As a consequence, the attractor at some point must fold over itself. Furthermore, even if the orbits diverge in different paths, sooner or later they have to pass close each other again. To visualize this process, we can think that the orbits located on a chaotic attractor are "mixed", just like a baker mixes the bread dough. The apparent randomness of the chaotic orbits is a consequence of this mixing process. Stretching and folding occur several times and produce folds and filaments within other folds and filaments, virtually endlessly. One can imagine what happens to nearby trajectories on a chaotic attractor by pouring a drop of black dye into the dough. Kneading is a combination of two actions: stretching of the dough, and folding of the dough onto itself. At first the drop of dye stretches, but then it is folded and after a while it will be stretched and folded many times.

Observing the dough at different magnifications, we will see that it is composed of many black and white layers that propagate into the mass to very small length scales.

This process can be reproduced by many numerical dynamical systems, which are very useful for understanding how the stretching and folding process can generate chaos and produce patterns in space and time. One of these dynamical systems is the so-called "Arnold's cat map" (or transformation; Arnold and Avez 1968). The map domain is the unit square ("mod 1" statement in Eq. 2.1 below). Importantly, a unit square is isomorphic to a two-dimensional torus. This means that there is a one-to-one mapping of each point on the unit square to each point on the surface of a torus. Imagine taking a sheet of paper and forming a tube. One of the dimensions of the sheet of paper is now an angle coordinate that is cyclic, going around the circumference of the tube. Imagine now that the sheet of paper is flexible and you can bend the tube around and connect the top of the tube with the bottom, forming a donut. The other dimension of the sheet of paper is now also an angle coordinate that is cyclic. In this way a flat sheet is converted into a torus. We will come back to this point in the next chapter when introducing another dynamical system that we will use as a prototypical mixing system.

One possible way to write the Arnold's cat map is

$$\begin{aligned} x_{n+1} &= x_n + y_n \bmod 1 \\ y_{n+1} &= x_n + 2y_n \bmod 1 \end{aligned} \tag{2.1}$$

The domain of this system is the portion of the plane confined to $x[0, 1]$ and y $[0, 1]$, equivalent to a square with a unit side. The map is a discrete dynamical system: unlike continuous dynamical systems (i.e. those defined by differential equations) in which the solutions are calculated by integration, a discrete dynamical system is solved by performing "iterations". This means that the system transports, at each iteration, each point of coordinates (x_n, y_n) to a new point with coordinates (x_{n+1}, y_{n+1}), i.e. $(x_n, y_n) \rightarrow (x_{n+1}, y_{n+1})$. Anytime a point exceeds the upper or lower limit of the map domain it re-enters the domain from the opposite side (toroidal approximation). The Arnold's cat map is a transformation that has the characteristic of conserving areas. In other words, the quantity of "matter" which is present at the beginning in the system does not vary over time: i.e. the system is conservative (e.g. McCauley 1993).

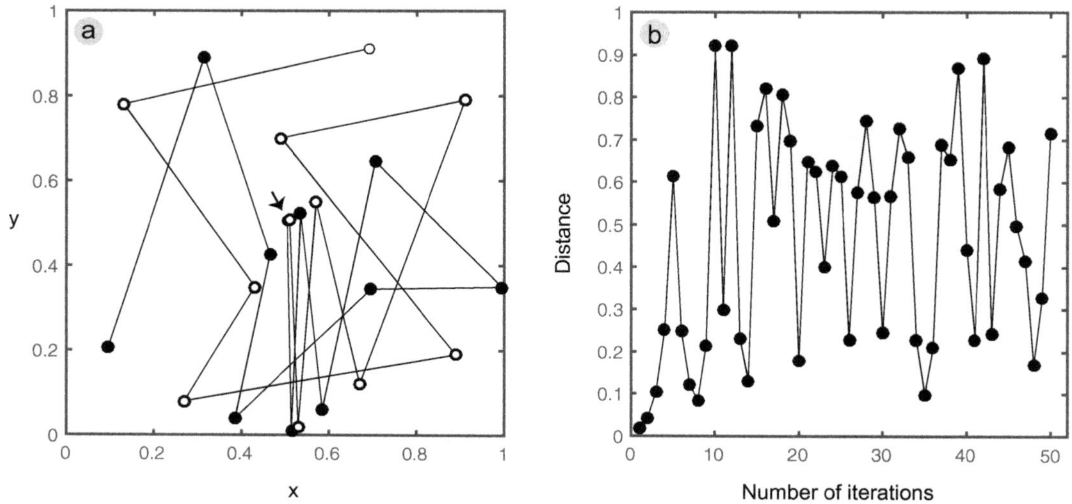

Fig. 2.4 **a** Paths described by two points initially placed very close each other (indicated by the black arrow) after they are iterated by the Arnold's cat map; **b** variation of the distance between the two points during the iteration of the map

Figure 2.4a shows the paths described by two points initially placed very close each other after they are iterated by the Arnold's cat map. The plot shows that after a very short time the two points move along completely different paths. This behavior highlights the sensitive dependence on initial conditions, typical of chaotic systems. The graph of Fig. 2.4b shows the variation of the distance between the two points during the iterations of the map. At the beginning the points move away from each other very quickly. However, given that the two points cannot diverge forever (given the finite size of the system) their distance decreases again. This behavior is repeated many times during the iteration of the map making the two points to "bounce" continuously in the domain of the map.

Figure 2.5 shows the action of the map on a "drop of dye" placed at the center of the map domain. As the number of iterations increases (from Fig. 2.5a to d), the drop is stretched and folded, producing increasingly thinner filaments that propagate inside the mixture, generating structures that are repeated at different length scales (Fig. 2.5e and f).

These structures show "scale invariance", that is, they recur with the same morphology (in this case filaments) at different magnifications. The generation of scale invariant patterns is a very important property. It is, indeed, a fundamental attribute of universal structures that are generated by chaotic systems, whose analysis allows us to quantifying chaos. These structures are "fractals".

2.6 Fractal Geometry

Western culture is obsessed with order and symmetry and we usually tend to impose on Nature patterns derived from classical Euclidean geometry. Despite the fact that Euclidean geometry is a gross simplification of the world, our society remained firmly devoted to this ordered view. This separated us progressively from Nature.

As children we are taught, for example, that natural objects such as trees, clouds or mountains can be represented using simple Euclidean constructs (e.g. triangles, circles, rectangles). Figure 2.6 shows an imaginary landscape drawn using only objects from Euclidean geometry. It is clearly a rough representation in which we can imagine to see natural shapes, but it is far from reproducing any real natural scenery.

A paradigmatic quotation from Mandelbrot (1982) supports very well these considerations: "Clouds are not spheres, mountains are not cones,

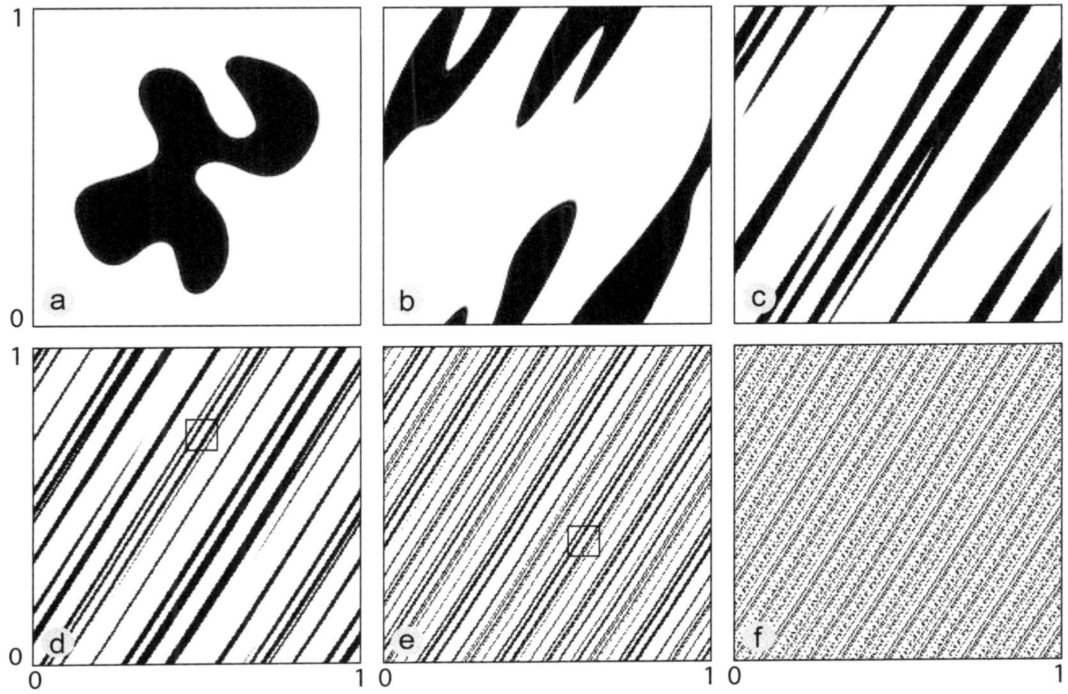

Fig. 2.5 Stretching and folding of the Arnold's cat map starting from the initial configuration shown in (**a**); as the stretching and folding progresses in time it produces progressively thinner filaments, generating structures that are repeated at different length scales (**e**–**f**). Black squares in (**d**) and (**e**) identify system portions which are magnified in (**e**) and (**f**), respectively

Fig. 2.6 Imaginary landscape drawn using only objects from Euclidean geometry

coastlines are not circles, and bark is not smooth, nor does lightning travel in a straight line". He realized that Nature and classical geometry did not agree in many circumstances and, around the 1970s, developed a new type of mathematics to describe and analyze the irregularity of patterns of the natural world. Mandelbrot (1982) coined a new name for these new geometric patterns: "fractals".

Any attempt to reduce a fractal into smaller parts results in the emergence of additional patterns, which in turn contain others and so on, just as in the case of the Arnold's cat map described in the previous section. This makes the objects described by fractal geometry profoundly different from those described by Euclidean geometry.

To better understand these differences, consider two objects: a circle, a classic Euclidean geometric shape and the coastline of Great Britain, a natural shape. What is the difference between the circle and the coastline? A useful technique for dealing with this type of problems is renormalization. It is a method of finding the infinitesimal boundary structure of an object or a self-similar process by repeatedly enlarging smaller parts of the whole. During this operation the circle loses its structure and turns to a simple straight line when it is observed at a very large scale. The circle is not a self-similar object. The case of the coastline of Great Britain is quite different. At different magnifications its boundary shows numerous particulars (headlands and bays) that propagate at different length scales. The coastline of Great Britain is a self-similar object: it is a fractal.

There are two classes of fractals: the natural and the mathematical ones. A coastline is a good example of a fractal shape that can be found in the natural kingdom. A mathematical curve with similar features is the Koch's curve (Mandelbrot 1982; Fig. 2.7). Here bays and headlands are increasingly smaller equilateral triangles. We cannot build a model of a coastline using the Koch's curve because Nature does not sculpt coastlines with equilateral triangles. However, the Koch's curve captures an important feature of natural coastlines: their scale-invariance. Natural and mathematical fractals not only display structures at all scales, they also display the same structures at all scales.

A portion of the coastline enlarged several times still resembles a coastline; the same holds for a segment of the Koch's curve. In the first case, however, the similarity is only statistical: the average proportions of bays and promontories remain the same at the different scales, although their precise arrangement may change. Typically, natural fractals exhibit self-similarity over a narrow range of scales. For example, the coastline is limited at the top by the largest geographical map that can represent it, while it is limited at the bottom by the most detailed map. In the second case, i.e. mathematical fractals, the self-similarity property propagates to infinity.

2.7 Fractal Dimension

How can we measure a fractal? Making quantitative measurements of all the individual details of a fractal is very difficult, but we can attempt to measure its level of irregularity. This measure, originally known as the "Hausdorff-Besicovitch dimension", is today commonly referred to as "fractal dimension" (Mandelbrot 1982). We are used to consider a line having one dimension, a plane two dimensions, a solid three dimensions. In the context of fractals, the dimension does not necessarily need to be an integer. The fractal dimension (generally denoted by the letter D) of a coastline, for example, is generally between 1.15 and 1.25 and that of the Koch's curve is close to 1.26 (Mandelbrot 1982). Thus, coastlines and the Koch's curve exhibit a similar level of irregularity. But what does it mean that the dimension of an object is fractional? The Koch's curve and the coastlines are obviously more rippled and more suitable for filling the space than a regular curve, which has dimension one, but less suitable for filling the plane, which has dimension two. Therefore, it is correct to say that their fractal dimension has to be between one and two. The fractal dimension is defined to capture this idea, and can be derived according to the same approaches used to define the dimension of traditional spaces. If we take an object embedded in a Euclidean space of dimension E and divide its linear size by r in each spatial direction, its measure (length, area, or volume) will increase according to

$$N = r^E \tag{2.2}$$

This process is illustrated in Fig. 2.8. For example, if we divide the line, the square and the cube by $r = 2$ we obtain $N = 2, 4$ and 8, respectively. If we divide by $r = 3$ we have that $N = 3, 9$ and 27. The Euclidean dimension E can be calculated from Eq. 2.2. Taking the logarithm of both sides we obtain that

$$\log(N) = E \log(r) \tag{2.3}$$

Solving for E

$$E = \frac{\log(N)}{\log(r)} \tag{2.4}$$

Using Eq. 2.4 we can verify that for the line, the square, and the cube the relative Euclidean dimension is one, two and three.

We can use the same approach to calculate the dimension of the Koch's curve (Fig. 2.7). In this

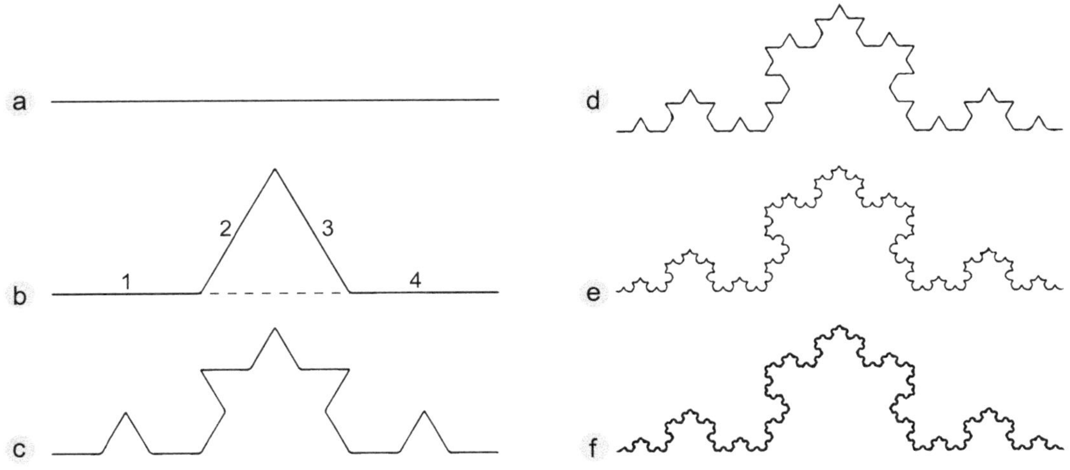

Fig. 2.7 Construction of the Koch's curve. The starting element is a segment of unit length (**a**). The segment is cut into three other segments and an equilateral triangle replaces the middle segment (**b**). The process is repeated recursively for each resulting segment (**c**–**f**)

Fig. 2.8 Illustration of a possible geometric method to derive Euclidean dimensions

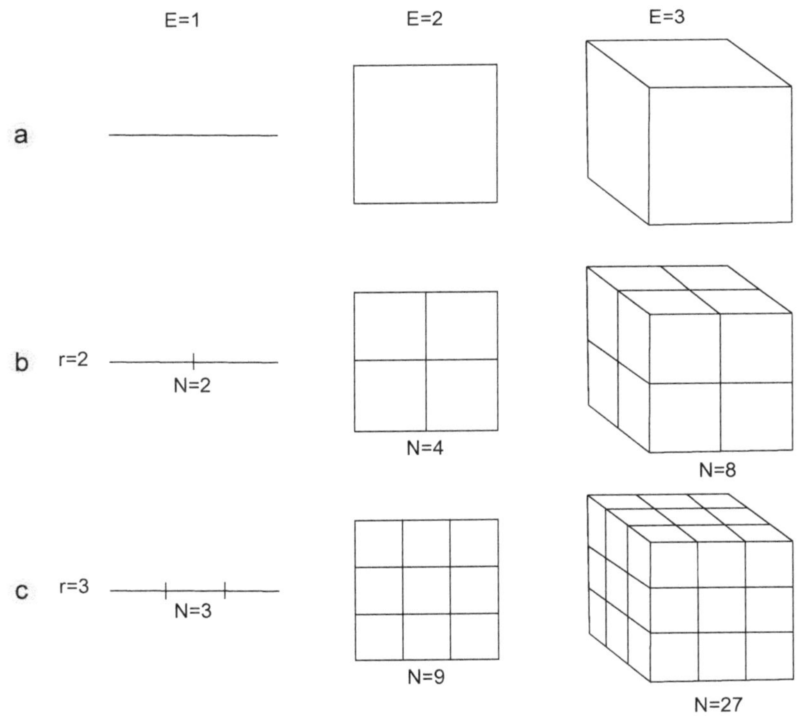

case the initial segment is divided by three ($r = 3$) obtaining $N = 4$. According to Eq. 2.4, the dimension of the Koch's curve is therefore ca. 1.26. As this dimension is not an integer, we call it fractal dimension and we indicate it with D ($D = 1.26$).

The method illustrated above to calculate fractal dimension can be used for mathematical fractals. However, it is not useful when dealing with natural patterns that do not obey to predefined mathematical rules. To overcome this difficulty,

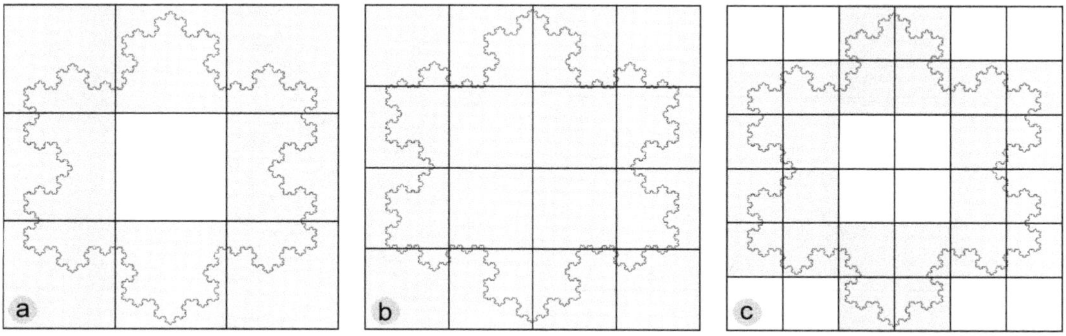

Fig. 2.9 Koch's snowflake built using the Koch's curve with a six-fold symmetry for which the box-counting technique is applied to estimate the fractal dimension

a number of techniques have been developed to estimate the fractal dimension of natural patterns. Among them, one of the most used is the so-called "box-counting" technique. Here we focus on this technique as it will be extensively used in the following chapters to measure the complexity of magma mixing patterns. Later in this chapter additional techniques to estimate the fractal dimension, which will be also used to quantify magma mixing, will be illustrated. The reader interested in a broader overview of other techniques for estimating the fractal dimension can refer to Kenkel and Walker (1996).

With the box-counting technique a square mesh of size (r) is laid over the pattern and the number of boxes (N_r) containing the pattern is counted (e.g. Mandelbrot 1982). For fractal patterns, the following relationship is satisfied:

$$N_r = r^{-D_{\text{box}}} \tag{2.5}$$

Using logarithms, Eq. 2.5 can be also written as

$$\log(N_r) = -D_{\text{box}} \cdot \log(r) \tag{2.6}$$

Equation 2.6 shows that, in order to classify a pattern as a fractal, data must lay on a straight line in the log-log space, where the fractal dimension (D_{box}) is estimated as the slope resulting from the linear fitting of the log(r) versus log(N_r) data.

Figure 2.9 shows the Koch's curve with a six-fold symmetry. Such a shape is commonly referred to as the Koch's snowflake. We can use this shape to estimate its fractal dimension with the box-counting technique.

The shape is covered with a grid of different box sizes (r) and the number of boxes (N_r) containing the pattern is counted. The plots of Fig. 2.10 show the variation of the box size (r) as a function of number of boxes (N_r) and the relative log-log plot. The graphs show that the relationship of Fig. 2.10a becomes linear in the log-log plot (Fig. 2.10b), as expected for a fractal pattern.

According to Eq. 2.6, linear fitting of the log-log plot allows us to estimate the fractal dimension (D_{box}) of the Koch's snowflake, which is equal to 1.258. This value is very close to the theoretical value of this geometric pattern ($D = 1.26$) that we have calculated before. The small difference between the calculated and estimated values is explained by the fact that in the first case we derived D analytically, whereas in the second case we estimated it using a finite resolution digital image. Nevertheless, the two values of fractal dimension agree very well.

2.8 Linking Chaos and Fractals

Chaos and fractals are two sides of the same coin. First, they both arise from the need to develop mathematical and physical models to describe natural processes and structures that cannot be studied by conventional methods. Second, in most cases chaotic dynamics generate fractal structures. Fractal geometry, therefore, offers the opportunity to study and quantify chaotic systems.

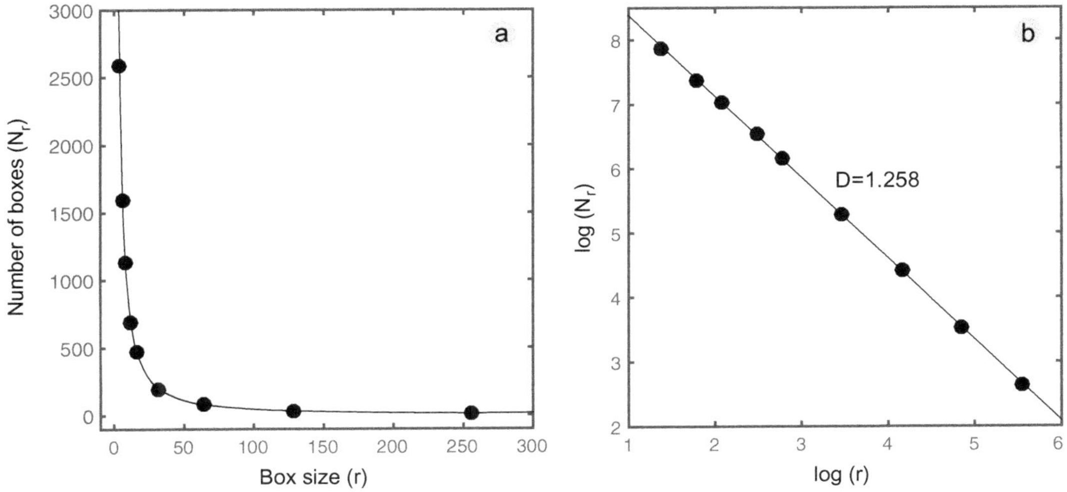

Fig. 2.10 Result of the application of the box-counting technique to the Koch's snowflake: **a** box-size (r) versus number of boxes (N_r) covering the shape; **b** log(r) versus log(N_r) showing the linear fitting of the data and the value of fractal dimension

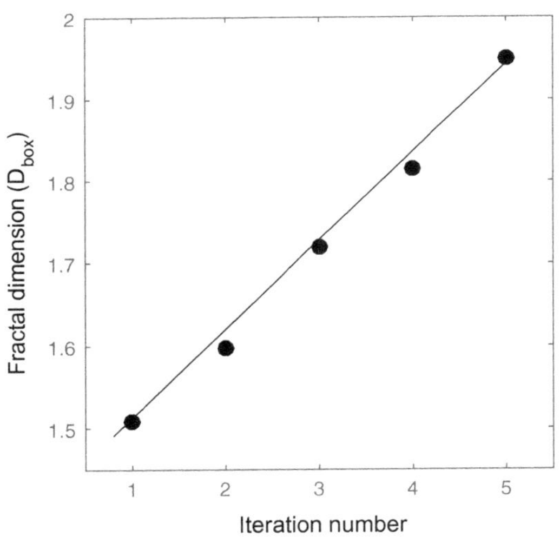

Fig. 2.11 Variation of fractal dimension for the patterns generated by the Arnold's cat map as the number of iterations of the map increases

As an example, this can be tested considering the Arnold's cat map introduced previously (Fig. 2.5). This dynamic system generates scale-invariant structures constituted by a large number of filaments that propagate within the domain of the map at many length scales. How to quantify the dispersion of the drop of dye as the stretching and folding develops in time? Since the patterns produced by the map show self-similarity, the simplest way is to measure their fractal dimension. We can do it using the box-counting technique. For example, D_{box} can be measured for the patterns produced by the map at different iterations.

The graph in Fig. 2.11 shows the number of iterations of the map as a function of fractal dimension. As the number of iterations increases, also the fractal dimension increases indicating that the fractal dimension is a suitable parameter to quantify the dispersion of the drop of dye in the mixture. More in general, since the process of stretching and folding within a system is the fundamental dynamics originating chaos, the fractal dimension is a measure that can be used to quantify the intensity of chaos in a dynamical system.

2.9 Further Methods to Estimate the Fractal Dimension

The box-counting method illustrated above can be conveniently used to measure the fractal dimension of digital images. However, when dealing with other types of natural structures such as compositional patterns developing in space or time, it might be useful to utilize other techniques. Hereafter, two methods that can be used to measure the fractal dimension of fragmentation processes and time and/or compositional series are described.

Fractal fragmentation

As reported by Mandelbrot (1982), Korcak (1940) performed empirical studies on the size distribution of the areas of islands and developed the empirical relationship:

$$N \approx a^{-c} \quad (2.7)$$

where N is the total number of islands having size greater than a given comparative size, a, and c is a constant (Korvin 1992). Mandelbrot (1982) found that c varied between island regions with c always being greater than $1/2$. In the light of fractal theory, he further realized that the size distribution of a population of islands was a consequence of fractal fragmentation and that the empirical constant c correlated with the fragmentation fractal dimension. He therefore suggested that fragmentation could be quantified by measuring the fractal dimension from cumulative distributions of fragments through the equation

$$N \approx r^{-D_f} \quad (2.8)$$

where D_f is the fragmentation fractal dimension and N is the total number of particles with linear dimension greater than a given comparative size, r. It is noteworthy that D_f derived from Eq. 2.8 is not a measure of irregularity, but a measure of the size-number relationship of the particle population or, in other terms, the fragmentation of the population. Taking the logarithm of both sides of Eq. 2.8 yields a linear relationship between N and r, with D_f being the slope coefficient:

$$\log(N) \approx -D_f \log(r) \quad (2.9)$$

The larger the value of D_f, the larger is the fragmentation efficiency.

It is interesting to relate Eq. 2.8 to a power-law relationship, which is extensively used as an empirical description for frequency-size distribution in the study of fragmentation processes (e.g. Turcotte 1992):

$$N \approx m^{-b} \quad (2.10)$$

where N is the number of fragments with mass greater than m. The constant b is chosen to fit the observed distribution. It can be shown that the constant b is equivalent to the fractal dimension D_f: since fragments can occur in a variety of shapes, it is appropriate to define a linear dimension r as the cube root of the volume

$$r \approx V^{1/3} \quad (2.11)$$

and, assuming constant density, it follows that $m \approx r^3$. Comparing Eq. 2.10 with the fractal distribution of Eq. 2.9 gives

$$D = -3b \quad (2.12)$$

This implies that the power-law distribution of Eq. 2.10 is equivalent to the fractal distribution of Eq. 2.8:

$$N \approx m^{-b} \approx m^{-D/3} \quad (2.13)$$

$$N \approx r^{-D} \quad (2.14)$$

Many size distributions in Nature follow this empirical law. As an example, it was shown that fragmentation of rock material is a consequence of the scale invariance of the fragmentation mechanism, in that the zones of weakness along which fragmentation occurs can be found at all levels of scrutiny (Turcotte 1992).

A paradigmatic model for fractal fragmentation is displayed in Fig. 2.12a, where two diagonally opposed blocks are retained at each scale. This corresponds to the comminution model proposed by Sammis et al. (1987).

The model is based on the hypothesis that direct contact between two fragments of near equal size during the fragmentation process will result in the breakup of one of the blocks. Then any initial particle distribution will evolve towards a distribution characterized by a minimum number of equal-sized particles at any scale. Furthermore, this model is based on the idea that it is unlikely that small fragments will break large fragments

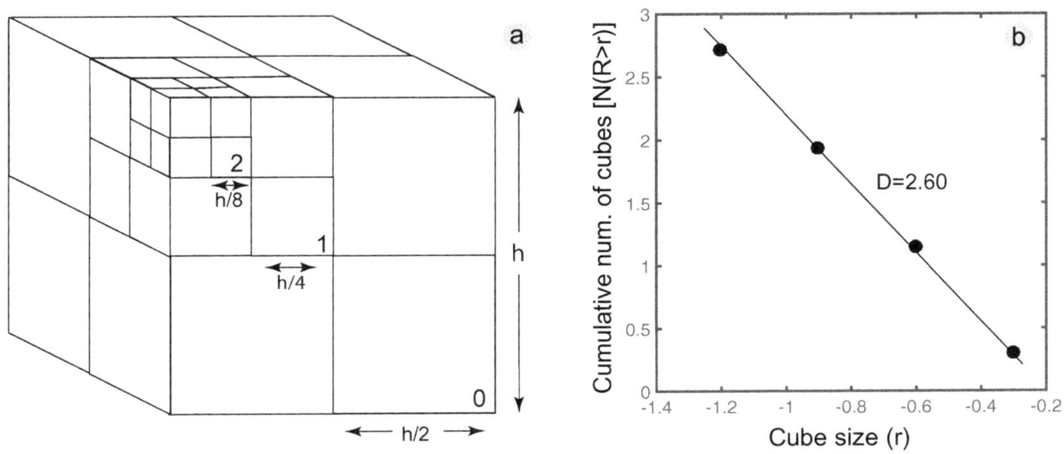

Fig. 2.12 **a** Idealized model for fractal fragmentation; **b** log(r) versus log[$N(R > r)$] plot showing the linear fitting of the data and the value of fractal dimension

or that large fragments will break small ones. According to the model shown in Fig. 2.12a, fragmentation starts from a cubic shape of size h and fragments into eight smaller cubes of size $h/2$. These smaller cubes are further fragmented following an iterative procedure to produce cubes with size $h/4$, and so forth.

Considering the cumulative statistics of fragments resulting from the application of the iterative procedure described above, the cumulative number of fragments larger than a specified size for the three highest orders are $N_{1c} = 2$ for $r = h/2, N_{2c} = 14$ for $r = h/4$, and $N_{3c} = 86$ for $r = h/8$; N_{Nc} is the cumulative number of the fragments equal to or larger than r_n, as required by Eq. 2.8. The cumulative statistics for the model illustrated in Fig. 2.12a are given in Fig. 2.12b. Applying Eq. 2.14, the fractal dimension of fragmentation for this model gives a value of $D_f = 2.601$. Such a value of D_f has been measured for a variety of rock types and appears to be a typical value for fragmentation of materials with brittle rheological behavior (Turcotte 1992).

Time series

The time evolution of a dynamic system can be represented by the time variation $X(t)$ of its dynamic variables. Any function $X(t)$ can be represented as the superposition of periodic components. The determination of their relative strengths is called spectral analysis.

If a time series $X(t)$ is specified over a time interval T, the mean signal $\bar{X}(t)$ is given by

$$\bar{X}(t) = \frac{1}{T}\int_0^1 X(t)dt \tag{2.15}$$

The variance σ^2, of the signal $X(t)$, is defined by

$$\sigma^2(X(t)) = \mathrm{Var}(X(t)) = \frac{1}{T}\int_0^1 [X(t) - \bar{X}(t)]^2 dt \tag{2.16}$$

and the standard deviation σ is the square root of the variance.

The mean and the variance are the first two moments of the time series. The time series σ^2, can be represented in the frequency domain f, in terms of the amplitude $A(f, T)$, which is the Fourier transform of $X(t)$:

$$A(f, T) = \int_{-\infty}^{\infty} X(t)e^{2\pi ift}dt \tag{2.17}$$

The inverse Fourier transform is

$$X(t) = \int_{-\infty}^{\infty} A(f, T)e^{-2\pi ift}df \tag{2.18}$$

Fig. 2.13 White noise time series (**a**) and its power spectrum (**b**) showing linear fitting of the data and the corresponding slope value (β)

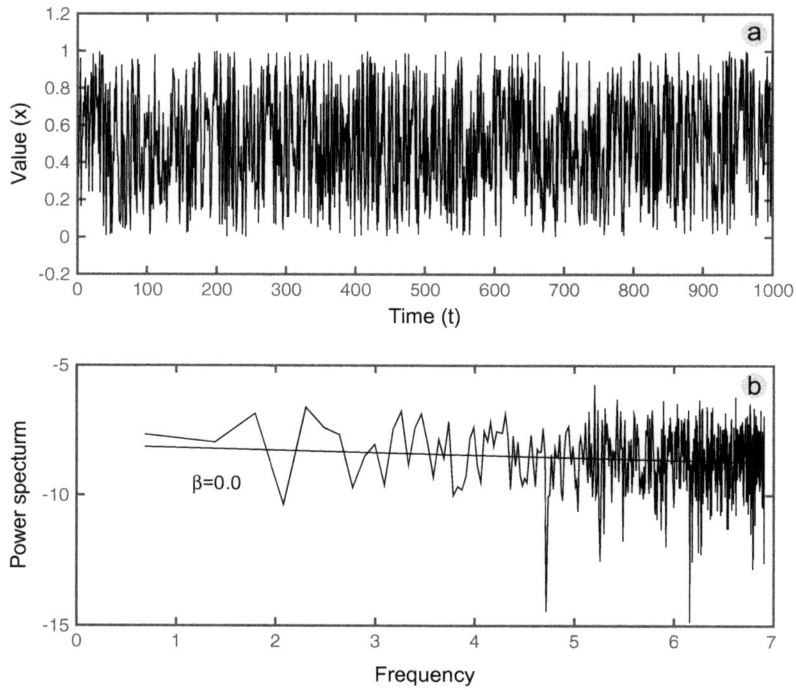

The quantity $|A(f, T)|^2$ is the contribution to the total energy of $X(t)$ from components with frequencies in $[f, f + df]$.

In real time series, the samples are picked up in a finite interval $0 < t < T$, in such a way that the effect of finite time series shall be taken into account. The FFT method (Fast Fourier Transform) is the appropriate tool to analyze this kind of systems (Brigham 1988). The power spectral density of $X(t)$ is defined by:

$$s(f) = \lim_{T \to \infty} \frac{1}{T} |A(f, T)|^2 \tag{2.19}$$

The quantity $s(f)df$ is the power in the time series associated with the frequency in the interval $[f, f + df]$. If a time series is fractal, then it satisfies the following power law relation:

$$s(f) \propto f^{-\beta} \tag{2.20}$$

where β is a constant that defines the kind of dynamic behavior of the time series $X(t)$. For example, $\beta = 0$ for white noise-like systems, which are uncorrelated and have a power spectrum that is independent of the frequency. Another relevant case is $\beta = 1$, the so-called flicker or $1/f$ noise, which is moderately correlated. For Brownian noise-like systems, $\beta = 2$, which are strongly correlated.

The spectral exponent β is linked with the fractal dimension D_{ts} of the time series. According to Higuchi (1988): when $1 < \beta < 3$, $D_{ts} = (5 - \beta)/2$; when $\beta \geq 3$, $D_{ts} = (7 - \beta)/2$; and when $\beta \leq 1$, $D_{ts} = (3 - \beta)/2$. Also, the following limits are held: if $\beta \to 0$ then $D_{ts} \to 2$; if $\beta \to 3$ then $D_{ts} \to 1$;

Power spectral analysis will be now applied to estimate the fractal dimension of three representative time series: the first is white-noise, the second is Brownian noise and the third one is the so-called $1/f$ flicker noise.

White noise is a random signal having equal intensity at different frequencies. In discrete time, white noise is a discrete signal whose samples are regarded as a sequence of serially uncorrelated random variables with zero mean and finite variance. Figure 2.13a and b display a typical white noise time series and the relative power spectrum. The power spectrum is flat according to the un-

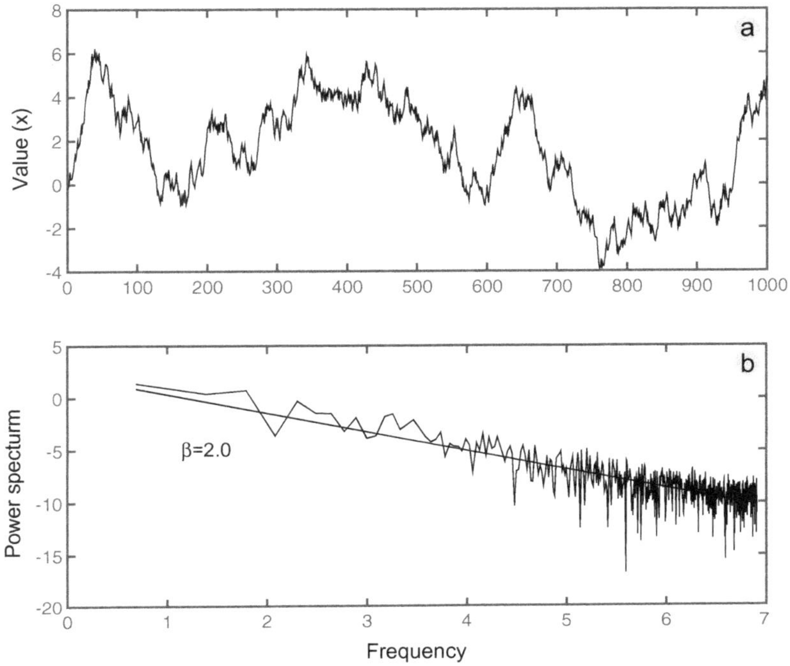

Fig. 2.14 Brown noise time series (**a**) and its power spectrum (**b**) showing linear fitting of the data and the corresponding slope value (β)

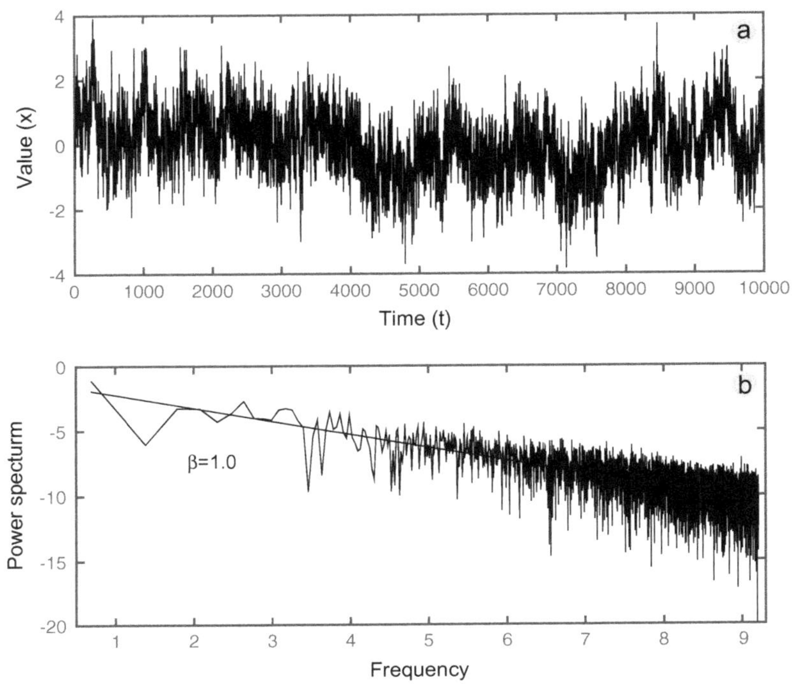

Fig. 2.15 $1/f$ flicker noise time series (**a**) and its power spectrum (**b**) showing linear fitting of the data and the corresponding slope value (β)

correlated nature of the time series. As a consequence, by applying the above relations, the slope of the power spectrum is $\beta = 0$ and the fractal dimension $D_{ts} = 2$.

Brownian noise, also known as Brown noise or red noise, is a signal noise produced by Brownian motion, hence its alternative name of random walk noise. The term "Brown noise" is used after Robert Brown, who documented the erratic motion for multiple types of inanimate particles in water. Brown noise can be produced by integrating white noise. In particular, while white noise can be produced by randomly choosing each sample independently, Brown noise can be produced by adding a random offset to each sample to obtain the next one. Figure 2.14a and b display a typical Brownian noise time series and its power spectrum. In this case, the slope of the power spectrum is $\beta = 2$ and, by applying the above relations, gives a value of fractal dimension $D_{ts} = 1.5$.

$1/f$ flicker noise, also known as pink noise, is a signal or process with a frequency spectrum such that the power spectral density is inversely proportional to the frequency of the signal. The name arises from the pink appearance of visible light with this power spectrum.

Flicker noise is characterized by correlations extended over a wide range of timescales. This is an indication of some sort of cooperative effect. Despite the ubiquity of flicker noise, its origin is not well understood. Flicker noise is, in fact, not noise but reflects the intrinsic dynamics of self-organized critical systems (Bak et al. 1988). We will come back to these concepts later in this book when dealing with plagioclase oscillatory zoning in magma mixing systems. Figure 2.15a and b display a typical flicker noise time series and its power spectrum. By applying the above relations, the slope of the power spectrum is $\beta = 1$ and the fractal dimension $D_{ts} = 1$.

References

Addison PS (1997) Fractals and chaos: an illustrated course. Institute of Physics Publishing

Alligood KT, Sauer TD, Yorke JA (1996) Chaos, an Introduction to dynamical systems. Textbooks in mathematical sciences. Springer New York, New York. ISBN: 978-0-387-94677-1. https://doi.org/10.1007/b97589

Arnold VI, Avez A (1968) Ergodic problems of classical mechanics. http://cds.cern.ch/record/1987366

Bak P, Tang C, Wiesenfeld K (1988) Self-organized criticality (Eng). Phys Rev A 38(1):364–374

Baker GL, Gollub JP (1996) Chaotic dynamics. Cambridge University Press, Cambridge. ISBN: 9780521471060. https://doi.org/10.1017/CBO9781139170864

Brigham EO (1988) The fast Fourier transform and its applications. Prentice Hall, Hoboken

Higuchi T (1988) Approach to an irregular time series on the basis of the fractal theory. Phys D: Nonlinear Phenom 31(2):277–283. ISSN: 01672789. https://doi.org/10.1016/0167-2789(88)90081-4

Kenkel NC, Walker DJ (1996) Fractals in the biological sciences. Coenoses 11:77–100

Korcak J (1940) Deux types fondamentaux de distribution statistique. Bulletin De l'Institute International de Statistique 3:295–299

Korvin G (1992) Fractal models in the earth sciences. Elsevier, Amsterdam

Lorenz E (1963) Deterministic non-periodic flow. J Atmos Sci 20:130–141

Mandelbrot BB (1982) The fractal geometry of nature. W.H. Freeman, New York

McCauley JL (1993) Chaos, dynamics and fractals: an algorithmic approach to deterministic chaos. Cambridge University Press, Cambridge

Sammis C, King G, Biegel R (1987) The kinematics of gouge deformation (English). Pure Appl Geophys 125(5):777–812

Turcotte DL (1992) Fractals and chaos in geology and geophysics. Cambridge University Press, Cambridge

3 The Chaotic Mixing of Fluids

The general idea of mixture is so familiar to us that the vast generalization to which these ideas afford the key, remains unnoticed.

O. Reynolds

Abstract

The study of chaotic mixing of fluids indicates that, even in simple geometries, fluids can show complex behaviours that could not be described by the deterministic dynamic systems of classical fluid mechanics. The mixing of fluids is the result of three main non-linear processes acting simultaneously and which are dependent on each other. They are stretching, folding, and molecular diffusion. The combination of stretching and folding triggers chaotic dynamics, generating intricate lamellar patterns that propagate in the system to very short length scales facilitating the homogenization process. These structures show scale invariance and are fractals. The formation of fractal lamellar structures implies an intimate contact between fluids, an exponential increase of contact area, and the development of compositional gradients, the latter triggering efficient diffusion processes. This chapter introduces the basic concepts of chaotic mixing of fluids to aid the reader to follow the arguments that will be presented in the next chapters about magma mixing processes.

3.1 Introduction

In the last forty years, the process of chaotic mixing of fluids has received considerable attention. Aref (1984) realized that, even in simple geometries, passive tracers dispersed in a host fluid may exhibit complex behaviours that could not be described by the deterministic dynamic systems of classical fluid mechanics. Since this pioneering work, a huge number of studies have been carried out. Some of them, such as Ottino (1990) and Aref and El-Naschie (1995), summarize the subject. The mixing of fluids is the result of three main non-linear processes acting simultaneously and which are dependent on each other. They are: (i) stretching, i.e. the elongation of the fluids, (ii) folding, i.e. the redistribution of fluids within the system, and (iii) molecular diffusion. We already encountered the stretching and folding couple in the previous chapter, where it was shown that this dynamical system constitutes the essence of chaos. For short mixing times, and if the fluids are

D. Perugini, *The Mixing of Magmas*, Advances in Volcanology,
https://doi.org/10.1007/978-3-030-81811-1_3

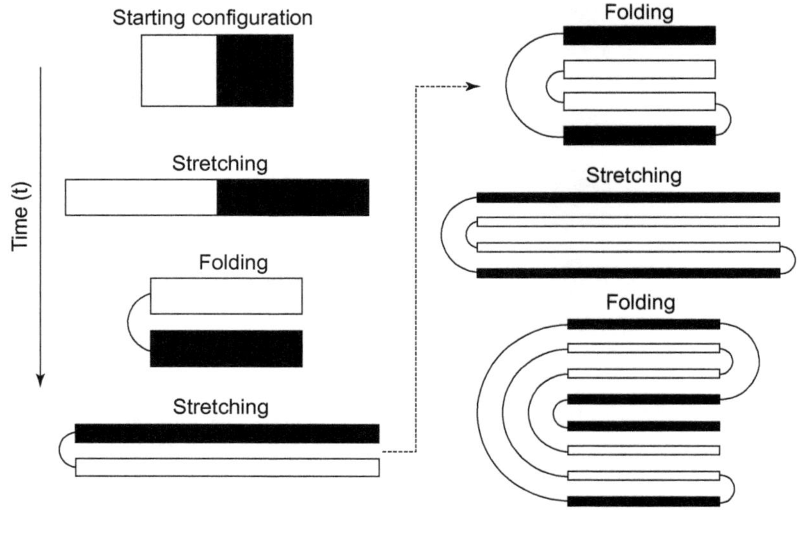

Fig. 3.1 Graphical representation of the Smale's horseshoe application

very viscous, molecular diffusion can be neglected. However, for longer mixing times this process can play a fundamental role in the hybridization of fluids. The structures and effects produced by the combined action of these three processes (stretching, folding and diffusion) have been extensively studied (e.g. Aref 1984; Aref and Balachandar 1986; Chaiken et al. 1986; Khakhar et al. 1986; Ottino 1990; Muzzio et al. 1992; Liu et al. 1994; Cerbelli et al. 2000; Anderson et al. 2000).

The general result is that the combination of stretching and folding triggers chaotic dynamics, generating intricate lamellar patterns that propagate in the system to very short length scales (e.g. Alvarez et al. 1998) facilitating the homogenization process. These structures show scale invariance and are fractals (e.g. Zumbrunnen et al. 1996). The formation of fractal lamellar structures implies an intimate contact between fluids, an exponential increase of contact area, and the development of compositional gradients that propagate at many length scales. This can induce molecular diffusion to erase the structures produced by stretching and folding dynamics.

The presence of chaotic dynamics in fluid mixing systems does not necessarily imply that the mixing process takes place efficiently. In fact, within a chaotic system different types of regions can coexist in which, depending on the type of attractor that locally manages the structure of the flow field, mixing occurs with different efficiency.

3.2 The Kinematics of Mixing

The kinematics of fluid mixing, although governed globally by the stretching and folding dynamics, is substantially related to the stretching process; folding is a consequence of the stretching process. In fact, given the finite size of the systems in which the mixing process takes place, the fluids cannot be stretched indefinitely. As a result, at some point the fluids must fold over themselves. From a strictly topological point of view, this operation corresponds to what in the literature is known as the Smale's "horseshoe" or Smale's "horseshoe application" (Smale 1967).

This topological transformation was already present in the Arnold's cat map described in the previous chapter and is graphically portrayed in Fig. 3.1. The repetition over time of Smale's horseshoes produces a dense set of lamellar structures. As mentioned above, the stretching and folding operation can generate, within the same mixing system, different types of regions whose dynamics are governed locally by different types of attractors. To illustrate in more detail the characteristics of these attractors, we consider a system whose Eulerian velocity field is $v(x, t)$.

The trajectory of an infinitesimal particle of fluid initially located in the position $x = x_0$ corresponds to the solution of the dynamic system:

$$\frac{dx}{dt} = v(x, t) \quad (3.1)$$

with the initial condition $x = x_0$. The solution of the differential equation (Eq. 3.1), for all x_0, is called the flow:

$$x = \phi_t(x_0) \quad (3.2)$$

In the case of a dynamic system in two dimensions $v(v_x, v_y)$ and $x(x, y)$. The flow corresponds to an application in the plane with the property of conservation of areas. The system has a Hamiltonian structure (e.g. Ottino et al. 1988) of the type:

$$\begin{aligned}\frac{dx}{dt} &= \frac{\partial \Psi}{\partial x}\\ \frac{dy}{dt} &= -\frac{\partial \Psi}{\partial y}\end{aligned} \quad (3.3)$$

where Ψ is the velocity field of the system (for Hamiltonian systems). If Ψ does not depend on time, the velocity field is integrable and the system cannot be chaotic. If Ψ depends on time, Hamiltonian mechanics indicates that the system can exhibit chaotic behaviors (e.g. Ottino et al. 1988). In the case of time-dependent velocity fields where $v(x, t) = v(x, t + T)$ and where T is the period, the flow described by Eq. 3.2 can be reduced to an application $x_n + 1 = \Phi T(x_n)$, where the position of x_0 at time $t = T$ is transformed into position x_1 at time $t = 2T$ and so forth; therefore, the application is iterated.

Assuming that the system in which the mixing occurs is closed, we can define an important property known as the "Poincaré recurrence" (e.g. Ottino et al. 1988). This property establishes that some fluid particles, after a certain period of time, will return exactly (or close to) to their original position (i.e. the particles "recur"). If a certain particle of fluid returns to its original position exactly after one period, i.e. $t = T$, the initial position corresponds to a periodic point of period one; if the particle returns after two periods, that is $t = 2T$, the point is of period two, and so forth. The superimposition of the images representing the positions of various initial conditions x_0 iterated for different times n_T, with $n = 1, 2, 3, ..., N$ for a large N shows the evolution of the Poincaré section of the initial conditions x_0. The Poincaré section represents the long-term evolution of the system for the considered initial conditions.

The key to understanding the complex behavior of fluid mixing lies in the characteristics of periodic points. They can be classified as elliptic, hyperbolic or parabolic, according to the deformation of the fluids in their proximity (Fig. 3.2). In general, the elliptical points are surrounded by regular islands ("elliptic sets") and, in this kinematic regime, the fluids trapped in these regions are rotated and weakly deformed, potentially retaining their identity for a long time (Fig. 3.2a). On the contrary, hyperbolic points are characterized by very strong deformations of the fluids due to the powerful stretching dynamics that occur within these regions ("hyperbolic sets"; e.g. Ottino et al. 1988; Fig. 3.2b). The parabolic points are exclusively associated with the presence of solid walls that geometrically delimit the systems and are characterized by weak translations of the fluids.

3.3 Iterated Maps as Prototypical Mixing Systems

In the study of mixing processes major problems occur because of the sensitive dependence on initial conditions typical of chaotic systems. This feature limits the number of techniques that can be utilized to perform numerical simulations since the prediction of the long-term evolution of the systems is denied. For instance, as reported by Metcalfe et al. (1995), using conventional integration procedures in tracking individual particle trajectories during numerical experiments, differences in the velocity field greater than 0.1% can produce, after only two or three circulation times, deviations of trajectories larger than the whole geometry of the system, making the computation useless.

To overcome such difficulties several other techniques have been proposed and in particular

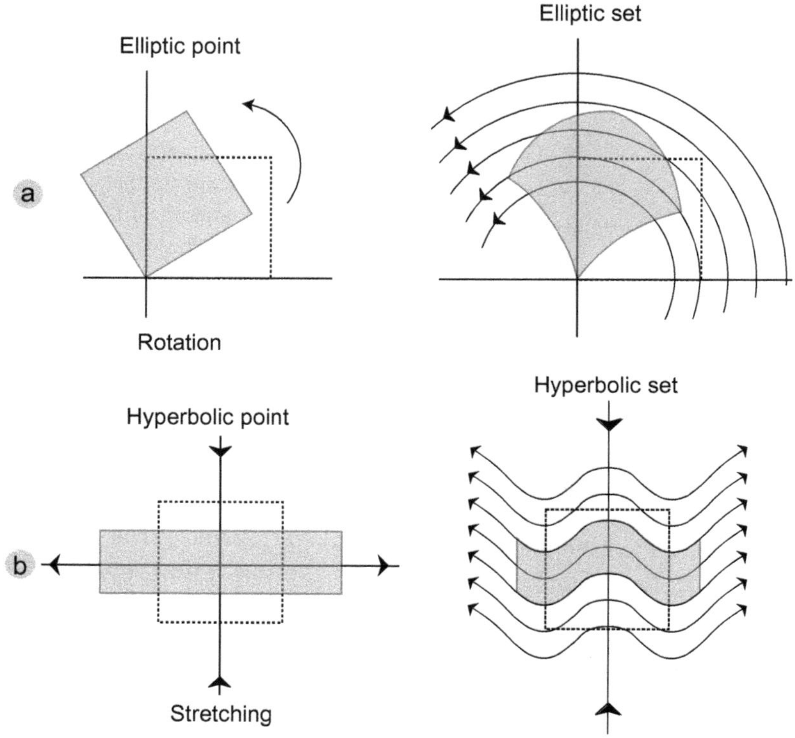

Fig. 3.2 Deformation of infinitesimal portions of a fluid in correspondence of an elliptic (**a**) and a hyperbolic (**b**) point. The corresponding right panels show the deformation of the fluid within an elliptical and a hyperbolic set. The hatched area represents the original shape before being deformed by the flow fields

the calculation of advection patterns has been recognized as one of the most suitable to simulate mixing processes (e.g. Metcalfe et al. 1995; Cerbelli et al. 2000). Underlying the concept of advection pattern is what is known as a "point transformation" (or "map"), a mathematical operation that enables to identify a particle of fluid and to specify its position at some time in the future. With this technique each fluid particle is "mapped" to a new position by the application of a transformation (or map) with extremely good precision avoiding the above-mentioned problems.

3.4 A Numerical Mixing Experiment Using Iterated Maps

In order to better understand the usefulness of iterated maps, we will now illustrate a numerical fluid mixing experiment. A similar system will be utilized in the next chapters when dealing with magma mixing processes. Here, however, we take into account only the stretching and folding process, neglecting molecular diffusion. Molecular diffusion will be investigated in detail in the case of magma mixing.

The iterated map considered here is known in the literature as the Chirikov-Taylor application or standard application ("standard map"; Chirikov 1979; Del-Castillo-Negrete et al. 1996). This dynamical system has the following formulation:

$$\begin{aligned} r_{n+1} &= r_n + K \sin(\theta_{n+1}) \quad \mod 2\pi \\ \theta_{n+1} &= \theta_n + r_n \quad \mod 2\pi \end{aligned} \tag{3.4}$$

The system is periodic in the interval $(\theta, r) \to [0, 2\pi]$ corresponding to a square with side equal to 2π (Fig. 3.3a). The position of each point $P(r_n, \theta_n)$ at time $t = n$ is transformed into a new position $P^1(r_{n+1}, \theta_{n+1})$ at time $t = n + 1$. K is a control parameter that can be varied to alter the flow field.

Since the system is periodic, if a point leaves the domain of the system, it is reintroduced from the opposite side (Fig. 3.3b). This operation

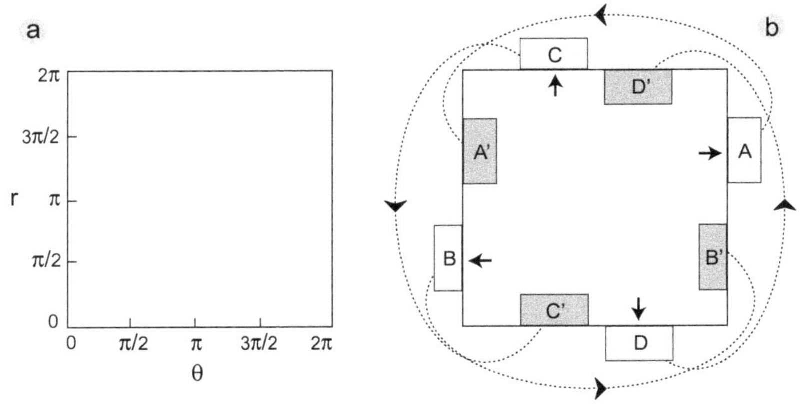

Fig. 3.3 Graphic representation of the domain of the Chiricov-Taylor application (**a**) and the principle of toroidal approximation used in the numerical simulations (**b**)

corresponds to what in the literature is known as a toroidal approximation of the domain of a dynamical system.

As mentioned above, the Poincarè sections contain information about the type of attractors embedded in the dynamical system. In order to construct the Poincarè sections of the Chirikov-Taylor map, we consider a random distribution of 300 points within the square of side $[0, 2\pi]$. Each point is then iterated 1000 times and its position is marked on the domain of the map. This procedure is repeated for different values of the control parameter K; the corresponding Poincarè sections are shown in Fig. 3.4.

For low values of K (Fig. 3.4a) regular regions (elliptic sets) almost completely dominate the dynamics of the system, leaving little space for chaotic regions (hyperbolic sets). As K increases, the regular regions decrease in size, and chaos takes over (Fig. 3.4b and c). Since hyperbolic sets (chaotic regions) are responsible for efficient stretching and folding dynamics K governs the mixing efficiency. The larger K, the better is the mixing. To understand the type of deformation that the fluids are subjected to within this dynamical system, we consider the system configuration for $K = 0.8$. A deformable grid is placed over the domain of the map (Fig. 3.5b). The Chirikov-Taylor map is iterated a number of times allowing the grid to deform according to the flow field.

Within the chaotic regions (hyperbolic sets), the mesh is strongly deformed highlighting the presence of powerful stretching and folding dynamics (Fig. 3.5c). On the contrary, within the regular regions (elliptical sets) the mesh is only slightly deformed, indicating that within these regions the fluids do not experience strong deformations (Fig. 3.5c).

To visualize in a more realistic way the impact of the map on fluid deformation, Fig. 3.6 shows the system filled by a white-colored fluid containing circular blobs of a black-colored fluid. One black blob is placed in the chaotic region, the other in the regular region. Figure 3.6a shows the initial configuration of the system. The evolution of the mixing process is shown in Fig. 3.6b and c after six and ten iterations, respectively.

The black fluid initially placed inside the regular region has not suffered appreciable deformation, in agreement with the fact that within these regions stretching and folding are very limited. On the contrary, within the chaotic regions, the black blob is violently stretched and folded, leading in a short time to lamellar structures. The fundamental point here is that within the same system, and at the same time, the two black blobs show very different mixing efficiency. In particular, Active Mixing Regions (AMR) and Coherent Regions (CR) coexist within the same dynamical system.

Figure 3.7 shows the growth of the interfacial area between the white-colored and the black-colored fluid for the numerical experiment presented in Fig. 3.6. The graph shows that, as the number of iterations increases, the black fluid inside the coherent region shows barely any

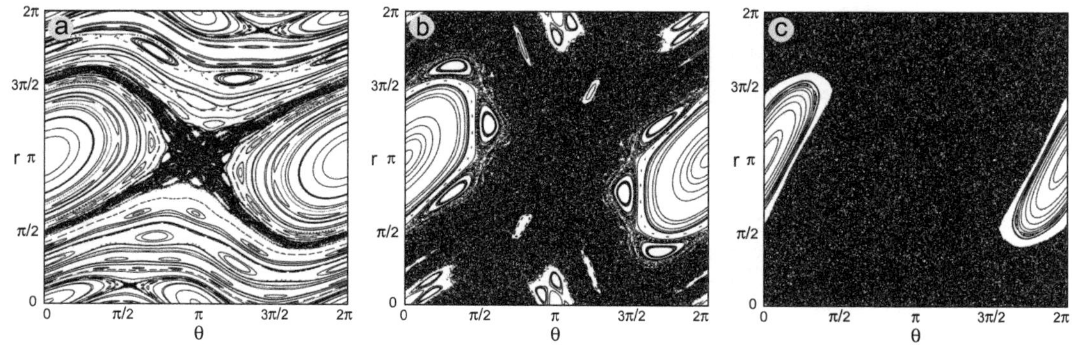

Fig. 3.4 Poincarè sections of the Chirichov-Taylor application for different values of the control parameter K: **a** $K = 0.8$; **b** $K = 1.5$; **c** $K = 3.0$

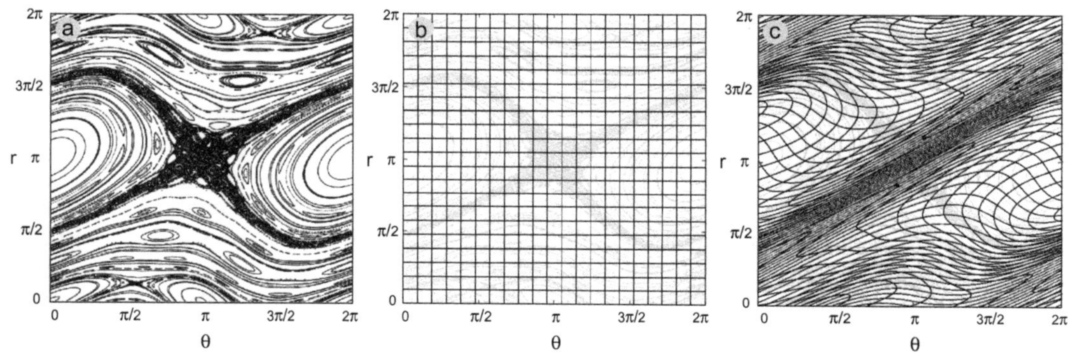

Fig. 3.5 A deformable grid **b** is placed over the domain of the Chirikov-Taylor map (**a**). The map is iterated allowing the grid to deform according to the flow field (**c**)

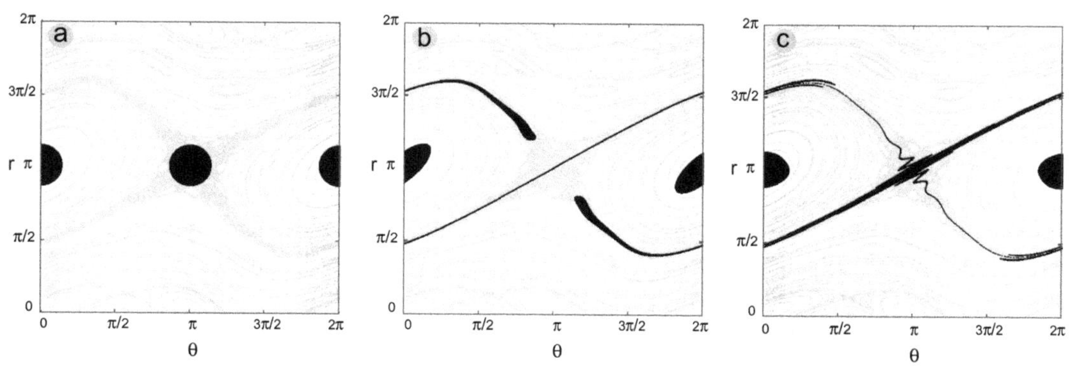

Fig. 3.6 Deformation of two blobs of a black coloured fluid in regular and chaotic regions

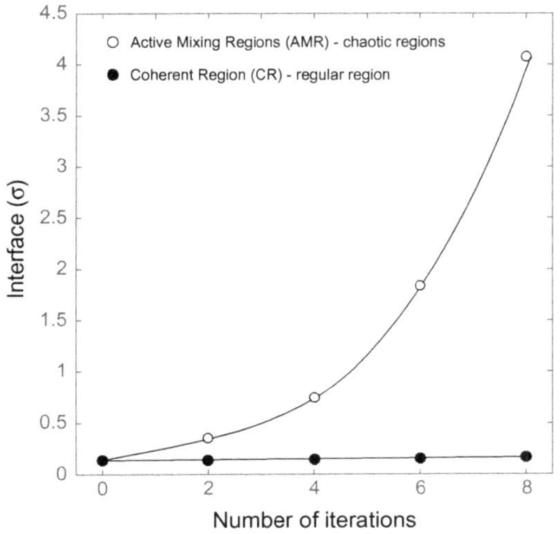

Fig. 3.7 Variation of the interfacial area between the white-colored fluid and the black-colored blobs in the regular and chaotic regions as a function of the number of iterations of the Chiricov-Taylor application

variation of the interfacial area. On the other hand, the black fluid within the chaotic region shows an exponential growth of the interfacial area.

The general result of this numerical experiment confirms that the chaotic regions are those zones of the dynamical system where the dynamics of stretching and folding act efficiently promoting the mutual dispersion of the two fluids. On the contrary, within the coherent regions the fluids can maintain their original identity for a long time.

The presence of regular islands (coherent regions) within mixing systems is usually considered a barrier to perfect mixing and, therefore, various solutions have been proposed to minimize the presence of elliptical regions (e.g. Liu et al. 1994; Lamberto et al. 1996). For example, in industrial applications, where the mixing process aims at being complete, regular islands are a plague for which a solution must necessarily be found (e.g. Liu et al. 1994; Lamberto et al. 1996). In the case of magma mixing processes, instead, the presence of regular regions is not that harmful. On the contrary, the presence of regular islands of a magma that have not been completely homogenized with the host magma constitutes the foundation to develop a new hypothesis for the interpretation of magmatic enclaves in both the plutonic and

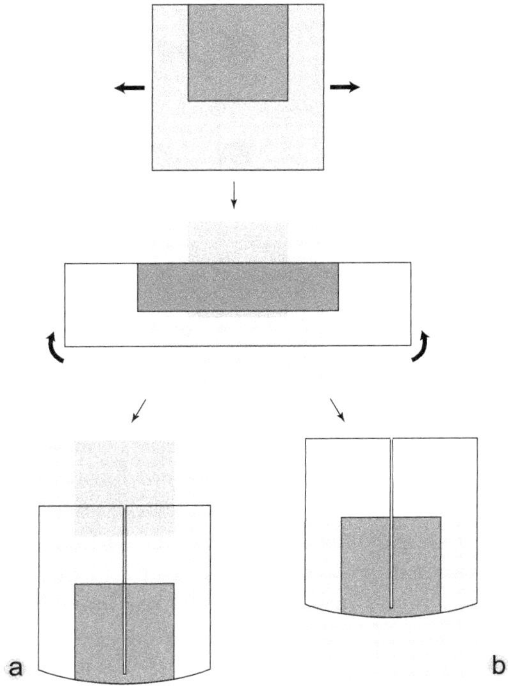

Fig. 3.8 Formation of regular regions. If the horseshoe is complete, the formation regular regions is precluded (**a**). If the horseshoe is incomplete, a regular region if formed. Modified from Bresler et al. (1997)

volcanic environment. Given the importance that regular regions can play in the study of the magma mixing, it is considered appropriate to discuss more in details the causes of their formation.

3.5 Regular Regions in an Ocean of Chaos

As mentioned above, the formation of regular regions is strictly related to the dynamics of stretching and folding processes. To illustrate in detail their formation, we consider again the Smale's horseshoe application. Figure 3.8 shows the genesis of a regular region during the development of stretching and folding. If stretching and folding generate a "complete" horseshoe (Bresler et al. 1997), the formation of regular regions is precluded. In fact, the portion of the dynamical system varies continuously in the domain of the system, and no fixed points are generated. This is

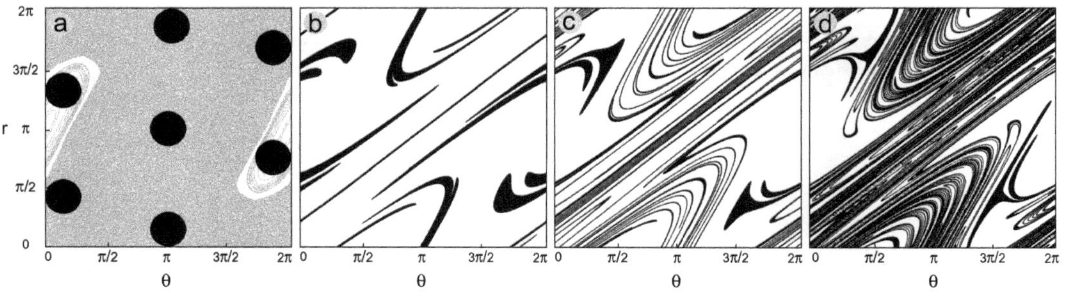

Fig. 3.9 Mixing between a white and black colored fluid simulated using the Chirikov-Taylor map

represented in Fig. 3.8a by the fact that the dark gray area, once the folding process is completed, resides outside the system area previously occupied. If, on the other hand, after the stretching and folding dynamics, the dark gray part is still in its original position ("incomplete" horseshoe; Bresler et al. 1997), its central part contains a fixed point that generates a portion of dynamic system that is spatially invariant (Fig. 3.8b). This necessarily evolves towards an elliptical set and a regular region is formed. Hence, elliptical regions are closely related to the topological structure of the flow field and are a fundamental aspect of any mixing system.

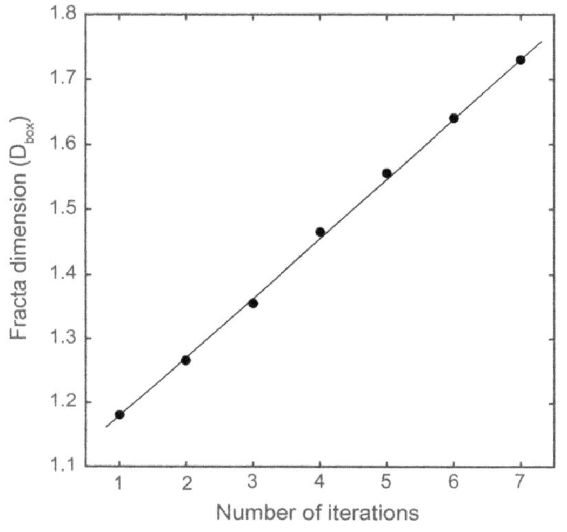

Fig. 3.10 Variation of fractal dimension (D_{box}) as a function of mixing time (i.e. the number of iterations of the map)

3.6 Fluid Mixing and Fractals

The repetition is space and time of stretching and folding dynamics in chaotic regions represents the cause for triggering efficient mixing processes. It is also the fundamental dynamics generating lamellar patterns that propagate into the system at many length scales, forming fractal domains.

This process is illustrated in Fig. 3.9, where the mixing process is simulated using the Chirikov-Taylor map starting from the initial configuration shown in Fig. 3.9a and for a value of the control parameter $K = 3.0$. With this value of K the system is mostly governed by chaotic regions and only one small regular region is present in the system. Figures 3.9b-d show the configuration of the mixing system after two, four and six iterations, respectively.

A possible way to quantify the mixing process is to use one of the fractal methods which we introduced in the previous chapter: the box-counting technique. The technique was applied to the picture produced after each iteration of the Chirikov-Taylor map and the relative value of fractal dimension (D_{box}) was estimated. The plot of Fig. 3.10 shows the variation of D_{box} as a function of mixing time (i.e. the number of iterations of the map). It is shown that fractal dimension grows linearly with mixing time, providing an effective method to measure the efficiency of the mixing process.

Before passing to the detailed study of magma mixing processes in the next chapters, a final consideration should be made on the use of iterated maps in the study of this natural process. Given the mathematical nature of the iterated map, such as the Chirikov-Taylor map, it could be possible, in principle, to follow the mixing patterns for an infinite number of iterations and magnifications. In the case of mixing between real fluids, since the structure of natural fractals is bounded between upper and lower limits (e.g. Mandelbrot 1982), it is not possible to observe lamellar structures at any length scale. In fact, the fractal nature of the magma mixing patterns can be appreciated only within a limited number of magnifications. For this reason, performing fractal analyses on natural magma mixing structures necessarily requires to carry out measurements at ca. the same scale. We will try to adhere to this point in the following chapters when dealing with estimates of the fractal dimension of magma mixing patterns.

References

Alvarez MM et al (1998) Self-similar spatiotemporal structure of intermaterial boundaries in chaotic flows. Phys Rev Lett 81(16):3395–3398

Anderson PD et al (2000) Mixing of non-Newtonian fluids in time-periodic cavity flows. J Non-Newtonian Fluid Mech 93(2-3):265–286

Aref H, El-Naschie MS (1995) Chaos applied to fluid mixing. Pergamon Press, Reprinted from Chaos, Solitions and Fractals, vol 4(6)

Aref H, Balachandar S (1986) Chaotic advection in a Stokes flow. Phys Fluids 29(11):3515–3521. ISSN: 00319171. https://doi.org/10.1063/1.865828

Aref H (1984) Stirring by chaotic advection. J Fluid Mech 143:1–21. ISSN: 14697645. https://doi.org/10.1017/S0022112084001233

Bresler L et al (1997) Isolated mixing regions: origin, robustness and control. Chem Eng Sci 52(10):1623–1636

Cerbelli S, Zalc JM, Muzzio FJ (2000) The evolution of material lines curvature in deterministic chaotic flows. Chem Eng Sci 55(2):363–371

Chaiken J et al (1986) Experimental study of lagrangian turbulence in a stokes flow. Proc R Soc Lond Ser A: Math Phys Sci 408(1834):165–174. ISSN: 00804630. https://doi.org/10.1098/rspa.1986.0115

Chirikov BV (1979) A universal instability of many-dimensional oscillator systems. https://doi.org/10.1016/0370-1573(79)90023-1

Del-Castillo-Negrete D, Greene JM, Morrison PJ (1996) Area preserving nontwist maps: Periodic orbits and transition to chaos. Phys D: Nonlinear Phenom 91(1–2):1–23. ISSN: 01672789. https://doi.org/10.1016/0167-2789(95)00257-X

Khakhar DV, Ottino JM (1986) Fluid mixing (stretching) by time periodic sequences for weak flows. Phys Fluids 29(11):3503. ISSN: 00319171. https://doi.org/10.1063/1.865824

Lamberto DJ et al (1996) Using time-dependent RPM to enhance mixing in stirred vessels. Chem Eng Sci 51(5):733–741. ISSN: 00092509. https://doi.org/10.1016/0009-2509(95)00203-0

Liu M, Muzzio FJ, Peskin RL (1994) Quantification of mixing in aperiodic chaotic flows. Chaos Solitons Fractals 4(6):869–893

Liu M et al (1994) Structure of the stretching field in chaotic cavity flows. AIChE J 40(8):1273–1286

Mandelbrot BB (1982) The fractal geometry of nature. W.H. Freeman

Metcalfe G, Bina CR, Ottino JM (1995) Kinematic considerations for mantle mixing. Geophys Res Lett 22(7):743–746

Muzzio FJ, Swanson PD, Ottino LM (1992) Mixing distributions produced by multiplicative stretching in chaotic flows. Int J Bifurcat Chaos 2:37–50

Ottino JM (1990) Mixing, chaotic advection, and turbulence. Ann Rev Fluid Mech 22:207–253

Ottino JM et al (1988) Morphological structures produced by mixing in chaotic flows. Nature 333(6172):419–425

Smale S (1967) Differentiable dynamical systems. Bull Am Math Soc 73:747–817

Zumbrunnen DA, Miles KC, Liu YH (1996) Autoprocessing of very fine-scale composite materials by chaotic mixing of melts. Compos Part A: Appl Sci Manufact 27(1):37–47. ISSN: 1359835X. https://doi.org/10.1016/1359-835X(95)00011-P

Part II

Numerical and Experimental Simulation of Magma Mixing

Numerical Models 4

Essentially, all models are wrong, but some are useful.

G. E. P. Box, N. R. Draper, "Empirical Model-Building and Response Surfaces"

Abstract

Numerical models allow us to gather new information on many natural processes that was not possible to obtain using only conventional field, theoretical and experimental approaches. In this chapter, simple $2D$ and $3D$ numerical models, based on iterated maps, are presented with the aim of defining a general numerical scheme containing the main building blocks of magma mixing processes. These numerical systems will be used as kinematic templates to study the topology of this process and will be used in the rest of this book as prototypical magma mixing systems.

4.1 The Meaning of Numerical Modeling

The great advances in computer technology since the 1980s have made possible to use mathematical models to gather new information on many natural processes that was not possible to obtain using only conventional field, theoretical and experimental approaches.

Generally, numerical models are considered appropriate if they can represent and reproduce a real problem. However, this is not always possible, or it is possible only to a certain approximation. Despite the huge developments in numerical models, they still have limitations in reproducing many physical phenomena, especially when applied to complex flows, including magma mixing. The main limitations are not only related to the structure of the numerical model itself, but also on the use of approximate input parameters, such as viscosity and its dependence on temperature or information on the geometry where the process develops. This is particularly relevant for magmatic systems for which most of the input parameters are known with a large degree of approximation.

A numerical model can be considered useful if it has the following properties: (a) the mathematical model has the capability to capture the main characteristics of the system, (b) it is sufficiently robust and capable of working without failing for a reasonable time, (c) the solution is accurate and can be proven analytically and/or experimentally, (d) the model is efficient within a reasonable computation time.

D. Perugini, *The Mixing of Magmas*, Advances in Volcanology,
https://doi.org/10.1007/978-3-030-81811-1_4

A large number of numerical models are today available; however, they should not be blindly applied. In particular, it is essential to examine whether the mathematical model is appropriate for the analysis of the physical phenomenon that it is intended to simulate. Common examples of inappropriate use of numerical models are: (a) use of existing software as a "black box", (b) inadequate setup of the initial and boundary conditions, and (c) unsatisfactory mathematical formulation for the reproduction of the physical phenomenon.

In the context of magmatic systems and, in particular magma mixing processes, the scarce knowledge of a large number of key parameters undermines our ability to validate satisfactorily most of numerical models. Among them: the modulation of rheological behavior of magmas in time and space as a function of temperature and composition, the exact knowledge of diffusivities for major and trace elements in the two magmas, the definition of the geometrical configuration in which the process takes place, and so forth.

At first sight, therefore, it may appear that the use of numerical models can add little to our knowledge about magma mixing processes. This is, indeed, true if we pretend to simulate with great precision the evolution of every possible variable of the magmatic system. If, on the contrary, our goal is to define a general scheme containing the main building blocks of the process, numerical models can aid in providing further insights on its complexity. This implies that models need to be designed to simulate the essential dynamics without pretending to perform a detailed reproduction of the process, which will be not possible give our approximate knowledge of the system and its variables.

As discussed in Chap. 3, a possible approach to the study of mixing processes is the use of iterated maps (e.g. the Chiricov–Taylor map). These are simplified dynamical systems that can capture the dynamics of fluid mixing and, therefore, can be also helpful in the context of magma mixing. In the following sections, two of these dynamical systems (the so-called "sine flow map" in $2D$ and "ABC flow" in $3D$) will be illustrated. Here we use iterated maps as kinematic templates to study the topology of this process and they will be used in the rest of this book as prototypical magma mixing systems.

4.2 Two-Dimensional Modeling

4.2.1 Stretching and Folding (Advection)

In the simulation of mingling of magmas in $2D$, a dynamical system consisting of repeated stretching and folding processes will be considered. The model is constituted by two non-linear equations (iterated map) (e.g. Collet and Eckman 1980; Sepulveda et al. 1989). As shown in Chap. 3, these systems are dynamic templates which are useful to investigate fluid mixing (e.g. Liu et al. 1994; Ott and Antonsen 1989). The fact that these numerical systems allow us to simulate mixing processes independently from the specific geometries in which the process develops makes them particularly suitable in the analysis of the patterns produced by magma mixing.

The iterated map that we take into account here is the "sine flow" map (e.g. Liu et al. 1994; Clifford et al. 1998; Clifford 1999). The flow is defined by two motions:

$$\begin{aligned} x_{n+1} &= x_n + \frac{k}{2}\sin(2\pi y_n) \quad [\mathrm{mod}\,1] \\ y_{n+1} &= y_n + \frac{k}{2}\sin(2\pi x_{n+1}) \quad [\mathrm{mod}\,1] \end{aligned} \tag{4.1}$$

The map is a two-dimensional conservative (area-preserving) chaotic system. The [mod 1] statement indicates that the domain of the system is periodic between 0 and 1 [i.e., $0 \le (x, y) \le 1$]. (x_n, y_n) and (x_{n+1}, y_{n+1}) are the coordinates of each fluid particle at time $t = n$ and $t = n + 1$, respectively, and k is the control parameter of the map, as for the case of the Chiricov–Taylor map illustrated in Chap. 3. Kinematically, the flow is given by the coupled action of two orthogonal motions having sinusoidal velocity profiles.

Many different velocity profiles can be utilized to reproduce chaotic systems (e.g. Aref and El-Naschie 1995). The sine flow map is utilized here because it is a well-studied chaotic system (e.g. Liu et al. 1994; Clifford et al. 1998; Clifford

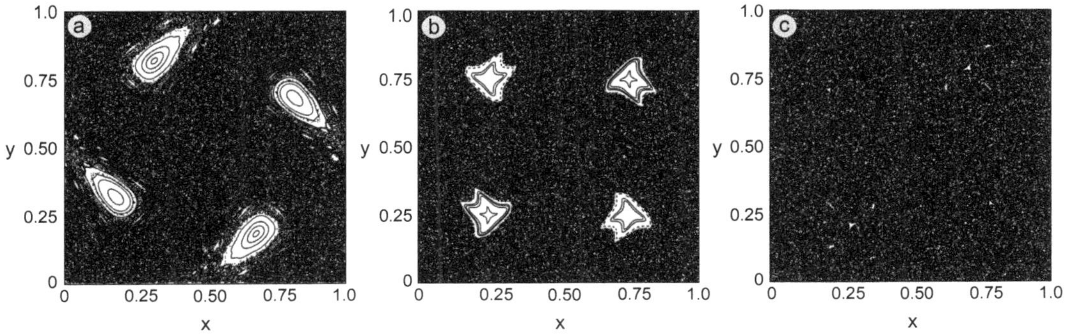

Fig. 4.1 Poincarè sections of the sine flow map for different values of the control parameter k; **a** $k = 0.4$, **b** $k = 0.5$, **c** $k = 0.6$

1999) and it has been successfully utilized to simulate magma mixing processes (e.g. Perugini et al. 2003, 2004).

The flow is defined on the $2D$ torus, as in the case of the Chiricov–Taylor application (Chap. 3), and different flow behaviours can be reproduced by varying the control parameter k (e.g. Liu et al. 1994; Clifford et al. 1998; Clifford 1999; Perugini et al. 2003, 2004). k is a parameter regulating the efficiency of advection, and is closely related to the degree and the efficiency of the mixing process. The possibility to explore different mixing intensities varying a single parameter (k) makes this map a useful prototypical system to study the complexity inherent to mixing processes.

Figure 4.1 shows the Poincar' sections of the sine flow map for different values of the control parameter k. For low values of k (Fig. 4.1a) the system is constituted by both regular (closed islands) and chaotic (dotted) regions. As k increases regular regions progressively disappear (Fig. 4.1b and c) allowing the chaotic region to extend over the entire domain of the system (Fig. 4.1c). Regular regions correspond to elliptic sets and are characterized by weak stretching dynamics compared to chaotic regions (hyperbolic sets). Therefore, the parameter k controls the relative presence of regular and chaotic regions in the mixing system. The larger k, the larger is the chaotic region and, consequently, the larger is the overall mixing efficiency of the system.

Quantifying the extent of stretching in these two different dynamical regions allows us to appreciate better the dynamic behavior of the numerical system. In particular, computations of stretching provide the means for characterizing the distribution of mixing intensities. Indeed, the positions of points experiencing high and low stretching determine regions of good and bad mixing. The stretching λ experienced by a segment after some time is defined as:

$$\lambda = |l| / |l_0| \tag{4.2}$$

The larger the value of stretching (λ) at a given position, the smaller the local striation thickness (Liu et al. 1994). Therefore, regions populated by points experiencing the largest and smallest stretching values correspond to regions of best and worst mixing, respectively. Figure 4.2 reports the calculation of the stretching field in the domain of the sine flow map for $k = 0.4$ and the corresponding probability density function.

Key features of the spatial structure of mixing are readily revealed by examination of the stretching field. In Fig. 4.2 the points are color coded according to the intensity of stretching. The color sequence, from lowest to highest stretching, is from blue to red. The stretching plot reveals that, in general, elliptic sets (i.e. closed islands, regular regions) are characterized by the lowest stretching values. Chaotic regions, instead, have the highest values of stretching. In addition, the stretching field reveals features of the mixing system to a surprising degree of detail. In particular, the highest stretching regions, although randomly distributed

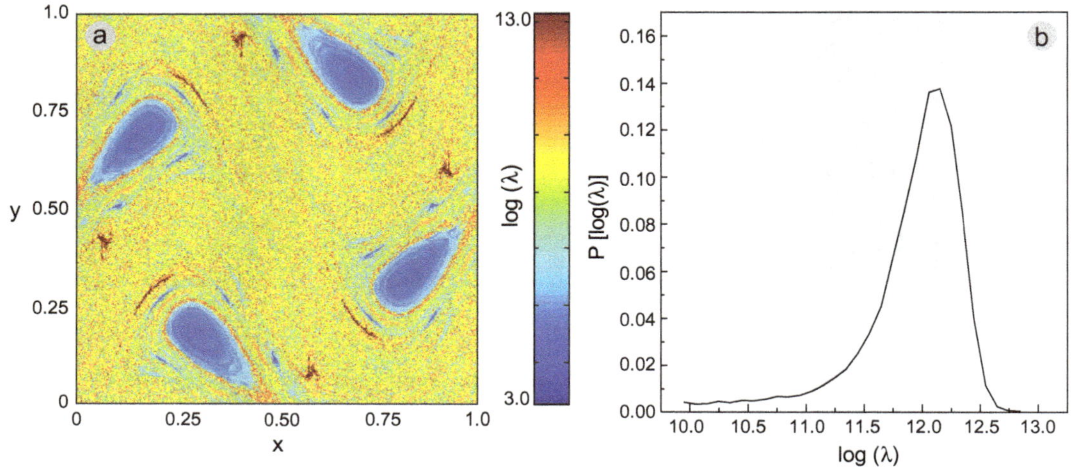

Fig. 4.2 **a** Distribution of stretching λ in the sine flow map domain for the value of the control parameter $k = 0.4$; **b** probability density function of the stretching field shown in (**a**)

in the chaotic domain, tend to cluster in spots around the regular regions. Here, they coexist with smaller islands of low stretching surrounding the regular regions. This effect would have been missed in Poincarè sections. For example, Poincarè sections would have captured the islands but not the low and high stretching regions around them. Therefore, the visualization of the stretching field allows us to penetrate into the complexity of the system to a greater detail.

The sine flow map can be used to trigger mixing processes and follow the development of mixing patterns in time. As an example, consider the initial configuration reported in Fig. 4.3a representing a distribution of black blobs of a magma (e.g. mafic) through a white colored (e.g. felsic) magma. Figure 4.3b, c and d show the configuration of the system after three, five and seven iterations (n) of the map for $k = 0.4$. The figure shows that the blobs initially placed into regular regions do not suffer strong deformation. On the contrary, they maintain their identity (with some minimal shape variations) during the time evolution of mixing. The case of the blob initially placed into the chaotic region is completely different. Here the liquid is efficiently stretched and folded, producing a huge number of filaments spreading throughout the entire chaotic region in a short time. The mixing system is, therefore, constituted by different dynamical regions in which the liquids suffer low and high deformations leading to a large heterogeneity of mixing efficiency.

A key point is the formation of lamellar patterns in the chaotic region that can propagate into the mixing system at many length scales, defining self-similar (i.e. fractal) patterns. As discussed in Chaps. 2 and 3, a possible method to measure the quality of mixing is to estimate the fractal dimension at different mixing times. Figure 4.4 reports the variation of fractal dimension (D_{box}) for numerical simulations of the mixing process using different values of the control parameter k. The initial configuration for all simulations is reported in Fig. 4.3a. The plot shows that, as the mixing process progresses in time, the fractal dimension (D_{box}) increases. For longer mixing times, D_{box} saturates to the value $D_{\mathrm{box}} = 2.0$, meaning that the mixing pattern progressively fills the $2D$ domain. Fitting of the linear portions of the plot for different values of k shows that the larger values of k correspond to larger values of the slope of the linear fitting. This is explained by the fact the larger k values correspond to a growth of the chaotic regions in the system and, therefore, to an increasing efficiency of stretching and folding dynamics to disperse the fluids.

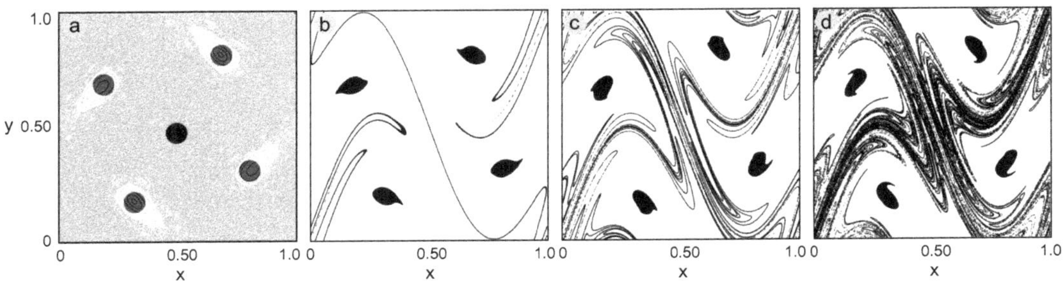

Fig. 4.3 Time progression of the mixing process (advection) in the sine flow map domain for the value of the control parameter $k = 0.4$. **a** Starting configuration of the simulation; **b–d** deformation of the blobs by the action of stretching and folding dynamics

4.2.2 Advection and Diffusion

Until now we considered the mixing process governed only by stretching and folding. This is, in fact, in the magma mixing jargon, what is usually referred to as "mingling". As reported in Chap. 1, hybridization (i.e. the generation of intermediate compositions between two initial liquids) cannot be achieve exclusively by mechanical mixing. Hybridization requires that stretching and folding dynamics must be combined with chemical diffusion.

To simulate chemical diffusion, we can define a continuous concentration field $c(x, y)$ on the plane. The concentration field is discretized into a number of cells [$c_{ij} = c(x_i, y_j)$; Fig. 4.5]. We consider the black colored magma constituted by elements having the same initial concentration equal to zero but different diffusion coefficients (D_{dif}). D_{dif} can vary from 0.0 to 1.0. The initial concentration of elements in the white colored magma is set to unity.

In the simulation of the diffusion process, diffusion coefficients are considered to be concentration-independent. The diffusion process is defined as:

$$c_{ij} = (1 - D_{\text{dif}}) \cdot c_{ij} + \frac{1}{4} \cdot D_{\text{dif}} \cdot \left[c_{(i+1)j} + c_{i(j+1)} + c_{(i-1)j} + c_{i(j-1)}\right] \quad (4.3)$$

where D_{dif} is the diffusion coefficient, c_{ij} is the concentration of a given cell and $c_{(i+1)j}$, $c_{i(j+1)}$, $c_{(i-1)j}$, $c_{i(j-1)}$ are the concentrations of the neighboring cells (Fig. 4.5). This numerical scheme is a typical finite difference scheme (e.g. Crank 1975; Pierrehumbert 1994) and corresponds to the discretization of the Laplace diffusion equation (e.g. Jain 1989). According to the diffusion that each cell experiences, c_{ij} is updated to new values ranging from 0.0 to 1.0. These two extreme values correspond to the original concentration of elements in the two magmas.

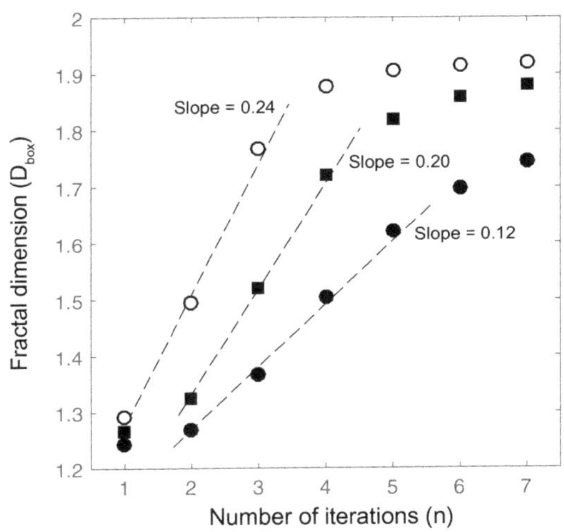

Fig. 4.4 Variation of fractal dimension (D_{box}) of mixing structures as a function of time and for different values of the control parameter k ($k = 0.4, 0.5, 0.6$). Slopes of the linear fitting, increasing with k, are also reported

Figures 4.6, 4.7 and 4.8 display the development of advection (i.e. stretching and folding) and diffusion between two hypothetical magmas, starting from the initial configuration reported in Fig. 4.6a and for values of the control parameter $k = 0.4$, $k = 0.5$, and $k = 0.6$, respectively. In the three figures the diffusion coefficient is kept

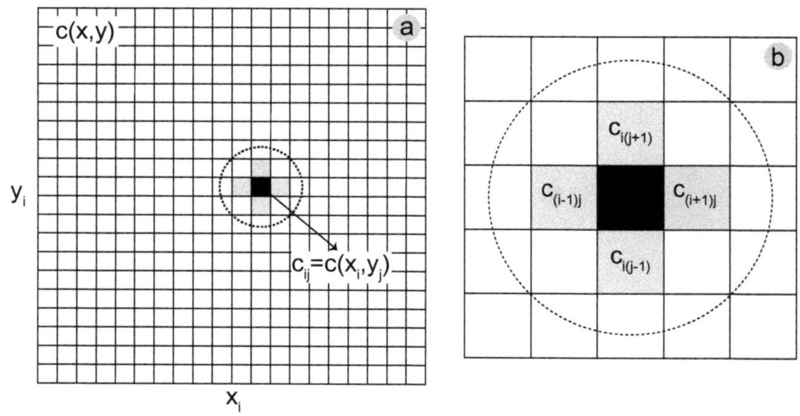

Fig. 4.5 **a** Discretization of the sine flow map domain for the simulation the chemical diffusion process; **b** magnification of the portion of dynamic system highlighted in (**a**) by the dashed circle showing the geometry of the neighborhood used to simulate diffusion

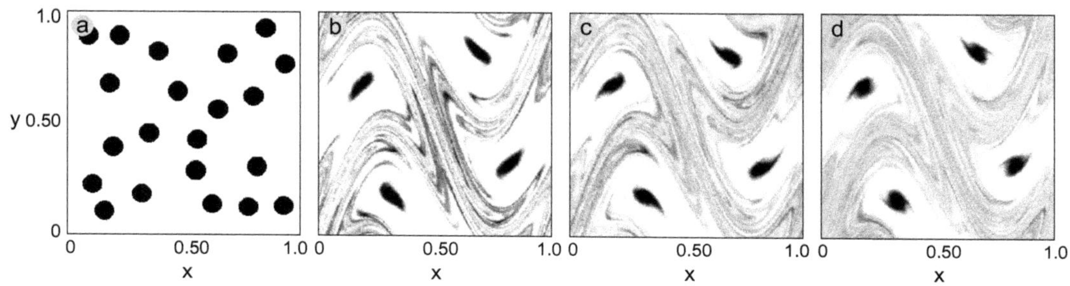

Fig. 4.6 Simulation of advection (stretching and folding) and diffusion carried out combining Eqs. 4.1 and 4.3 for different mixing times (n). The starting configuration of the simulation is given in (**a**). The value of control parameter is $k = 0.4$

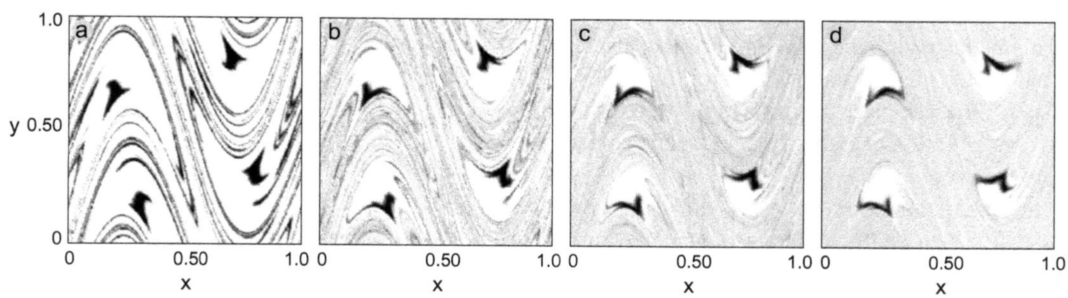

Fig. 4.7 Simulation of advection (stretching and folding) and diffusion performed combining Eqs. 4.1 and 4.3 for different mixing times (n). The starting configuration of the simulation is given in Fig. 4.6a. The value of control parameter is $k = 0.5$

constant to $D_{\text{dif}} = 0.8$. After some iterations, the lamellar patterns generated in the chaotic regions have almost completely disappeared and a fairly homogeneous region is generated. After the same number of iterations, the blobs of black magma originally placed in the regular regions are still recognizable (Fig. 4.6). This is a consequence of the fact that in the chaotic regions the two liquids are brought into contact along large contact interfaces. These interfaces propagate into the

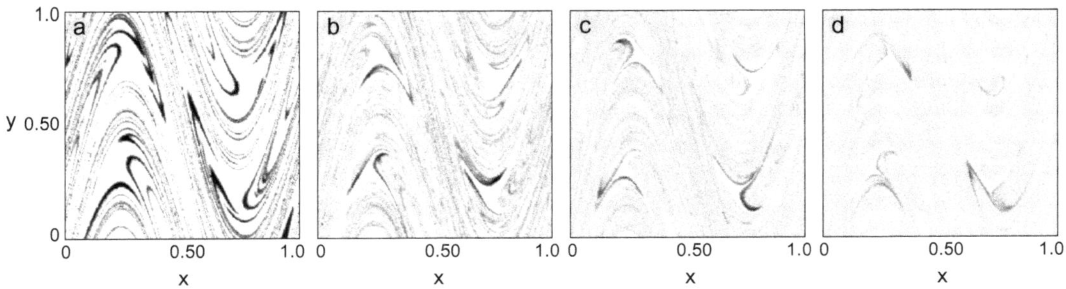

Fig. 4.8 Simulation of advection (stretching and folding) and diffusion performed combining Eqs. 4.1 and 4.3 for different mixing times (n). The starting configuration of the simulation is given in Fig. 4.6a. The value of control parameter is $k = 0.6$

mixing system at several length-scales. In these conditions chemical diffusion can easily "erase" them, producing hybrid compositions. On the contrary, when blobs of the black colored magma are trapped within regular islands, the contact interface with the white colored magma does not change greatly, prohibiting chemical diffusion to homogenize them with the host magma.

4.2.3 Concentration Variance

We have seen before that fractal dimension is a good measure of the mechanical mixing between fluids. However, in the presence of chemical diffusion this method may pose serious problems. Indeed, in order to apply the box-counting technique, black and white images are needed. If chemical diffusion occurred between the fluids, images will contain different shades of gray (i.e. different degree of hybridization). In this case, image reduction to black and white colors, may potentially lead to loosing information about the mixing process.

A useful measure in the study of mixing (intended as the combined action of advection and diffusion) should be able to quantify also the diffusion process. A quantity commonly used in the fluid dynamics literature to evaluate the degree of homogenization of fluid mixtures is the concentration variance (σ^2) (e.g. Rothstein et al. 1999; Liu and Haller 2004). The variance of concentration for a given chemical element (C_i) is given by

$$\sigma^2(C_i) = \frac{\sum_{i=1}^{N}(C_i - \mu_i)^2}{N} \tag{4.4}$$

where N is the number of samples, C_i is the concentration of element i and μ is the mean composition.

Concentration variance is destined to decrease with mixing time because the system becomes progressively more homogeneous. Variance defined by Eq. 4.4 depends on absolute values of concentrations of chemical elements. In order to avoid this, σ^2 is rescaled to the initial variance $[\sigma^2(C_i)_{t=0}]$. Hereafter, we will refer to concentration variance considering the following quantity:

$$\sigma_n^2 = \frac{\sigma^2(C_i)_t}{\sigma^2(C_i)_{t=0}} \tag{4.5}$$

where $\sigma^2(C_i)_t$ and $\sigma^2(C_i)_{t=0}$ is the concentration variance (C_i) at time t and time $t = 0$ (i.e. the initial variance before the mixing starts), respectively. This measure quantifies the degree of homogeneity of the mixing system. In detail, the concentration variance σ_n^2 varies between unity at $t = 0$ and zero at $t = \infty$ (i.e. the time at which the system is homogeneous).

Figure 4.9 shows the variation of concentration variance as a function of the number of iterations of the sine flow map for three hypothetical chemical elements having $D_{\text{dif}} = 0.05, 0.1, 1.0$. As the black colored magma is stretched and folded in time, chemical diffusion becomes progressively

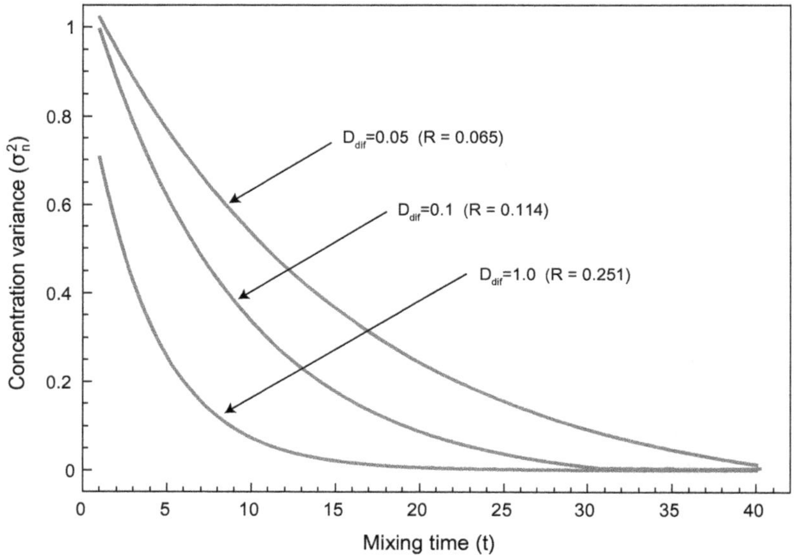

Fig. 4.9 Decay of concentration variance with the time progression of the mixing process for three chemical elements having different values of the diffusion coefficient (D_{dif}). Values of the rate of concentration decay (R) are also reported

more efficient due to the generation of new interfacial areas. As a consequence, the system becomes gradually more homogeneous and, hence, concentration variance (σ_n^2) decays in time. σ_n^2 decreases quickly during the first stages of mixing and then shows a relaxation toward $\sigma_n^2 = 0$ for long mixing times, as the system approaches homogeneity.

Several works (e.g. Rothstein et al. 1999; Mathew et al. 2007) reported that during chaotic mixing the change of concentration variance can be fitted by an exponential function of the form

$$\sigma_n^2 (C_i) = K_0 \cdot \exp(-Rt) + K_1 \qquad (4.6)$$

where K_0, R and K_1 are fitting parameters and t is the mixing time. Equation 4.6 can be used to quantify the rate of decay of concentration variance σ_n^2 of the simulated chemical elements. The plot of Fig. 4.9 shows the fitting for the three chemical elements and indicates that, indeed, variance decays exponentially with mixing time. The rate of concentration decay is governed by the parameter R (Eq. 4.6). We call this parameter "Relaxation of Concentration Variance" (RCV) and we use it as a measure of the element mobility during mixing. Results from the fitting indicates that RCV values are different for the different elements (Fig. 4.9), with RCV values increasing as the diffusion coefficient (D_{dif}) increases.

4.2.4 Mixing and Entropy

In general terms, mixing indicates a process that reduces compositional heterogeneity. Entropy is the measure of disorder or system homogeneity. We will now explore the use of Shannon entropy (Shannon et al. 1948) as an additional measure to quantify the efficiency of mixing. In particular, the Shannon entropy can be applied to measure the intensity of mixing and its variation in time.

In order to apply this concept, numerical simulations of advection-diffusion with the sine flow map were performed using different values of the diffusion coefficient (D_{dif}). This time, however, the control parameter k is not kept constant during the simulations, but is varied randomly after each iteration of the map. Figure 4.10 shows the evolution of the mixing system starting from the initial configuration reported in Fig. 4.10a for a value of $D_{dif} = 0.3$. The figure shows, as expected, that the system becomes progressively more homogeneous as the mixing process progresses in

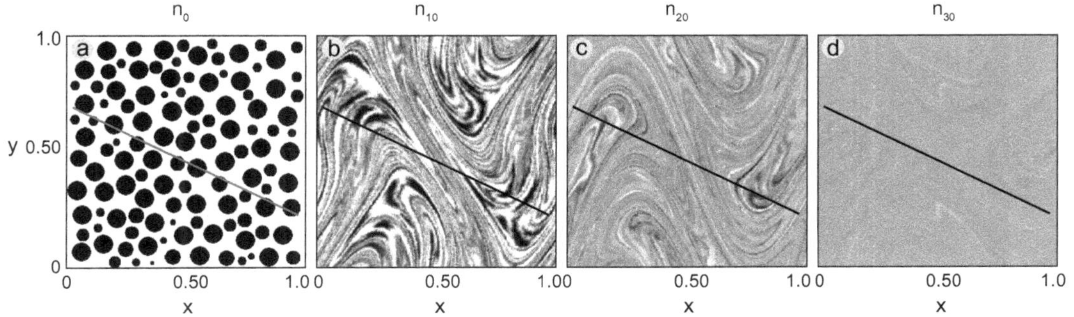

Fig. 4.10 Simulation of advection (stretching and folding) and diffusion performed combining Eqs. 4.1 and 4.3 for different mixing times (t). The starting configuration of the simulation is given in (**a**). During the simulation, the control parameter k is varied randomly after each iteration of the sine flow map

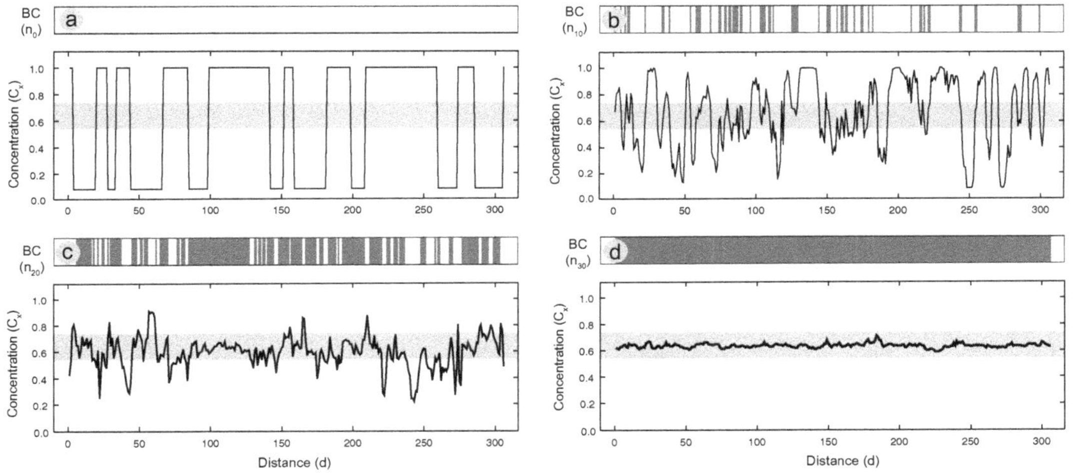

Fig. 4.11 **a–d** Compositional variation along a transect (Fig. 4.10) of the numerically simulated mixing system for an element with diffusion coefficient $D_{\text{dif}} = 0.3$ at different times [from (**a**) to (**c**)]. On the top of each compositional transect is reported, as a "bar-code", the number of samples having the hybrid composition in the mixing system. A tolerance of 10% is introduced in defining the hybrid composition

time. At each iteration of the sine flow map, the variation of concentration is monitored along transects crossing the mixing system (Fig. 4.10b–d). Although, in principle, the analyses reported below can be made for the whole system, here we prefer to focus on transects, as this kind of analysis is typically used in the study of natural samples (see next Chapters).

The numerical scheme was repeated for one hundred iterations. Figure 4.11 shows the change of concentration across the transect at different times. It is shown that the chemical oscillations are progressively smoothed out as the mixing time increases. For long mixing times, the compositional spectrum reduces to a single value representing the hybrid composition (Fig. 4.11d).

Figure 4.12 shows the tenth iteration of the mixing system for two elements having diffusion coefficient (D_{dif}) equal to 0.3 and 1.0. The pictures on the left indicate that, at the same time, the mixing system becomes progressively more homogeneous due to the increasing diffusion of elements. The graphs of the right show the compositional variability of the elements. The different elements display extremely different patterns. In detail, the slow diffusing element varies

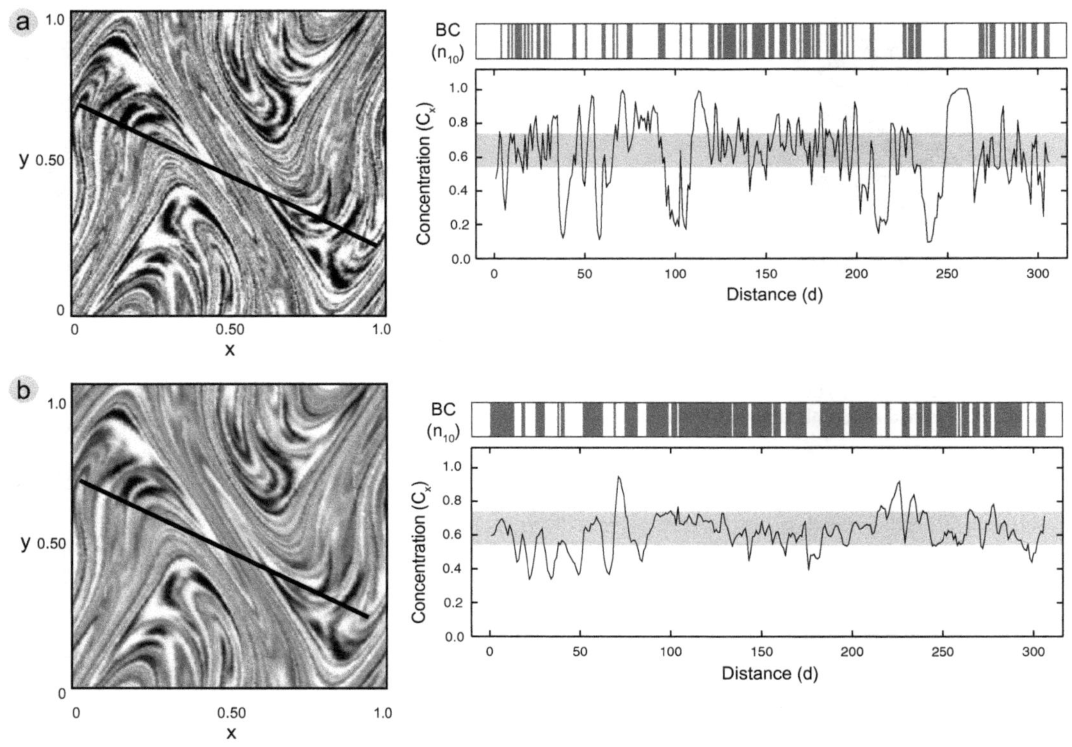

Fig. 4.12 Variability of composition in the mixing domain at the same number of iterations ($n = 10$) for elements with different D_{dif}. Left panels: mixing domain; black line indicates the transect used to monitor the compositional variability; right panels: variation of element concentrations along the transects reported on the left; the "bar-code" representation of the amount of the hybrid composition is also shown on the top of compositional transects

between compositions similar to those of the two end-members. Here, only a few data points are shifted towards the hybrid composition. Increasing D_{dif} values corresponds to patterns that are gradually less variable, generating an increase of the number of data points corresponding to the hybrid composition (Fig. 4.12). It is therefore clear that at the same time and in the same system chemical elements can show very different behaviour depending on their diffusivity. Slow diffusing elements can show a large variability; others appear completely homogenized.

The number of data points having the hybrid composition is monitored for each compositional series. For each sample that reached the hybrid composition, a vertical segment of arbitrary length is drawn along a line at the relative position of the data point. This produces a "bar-code" representation of the distribution of hybrid samples along the compositional transect (Figs. 4.11 and 4.12).

The hybrid sample distributions (compositional "bar-codes") extracted from the mixing system were used to calculate the evolution of Shannon entropy (S) (Camesasca et al. 2006; Shannon et al. 1948) of the mixing system in time. In particular, we consider the transect to be divided into a grid made up of a number M of cells c_i with area a_i, such that

$$\sum_{i=1}^{M} a_i = A \tag{4.7}$$

where A is the total area of the transect. As shown by Baranger et al. (2002), entropy (S) can then be calculated as

$$S(t) = -\sum_{i=l}^{M} p_i(t) \log p_i(t) \tag{4.8}$$

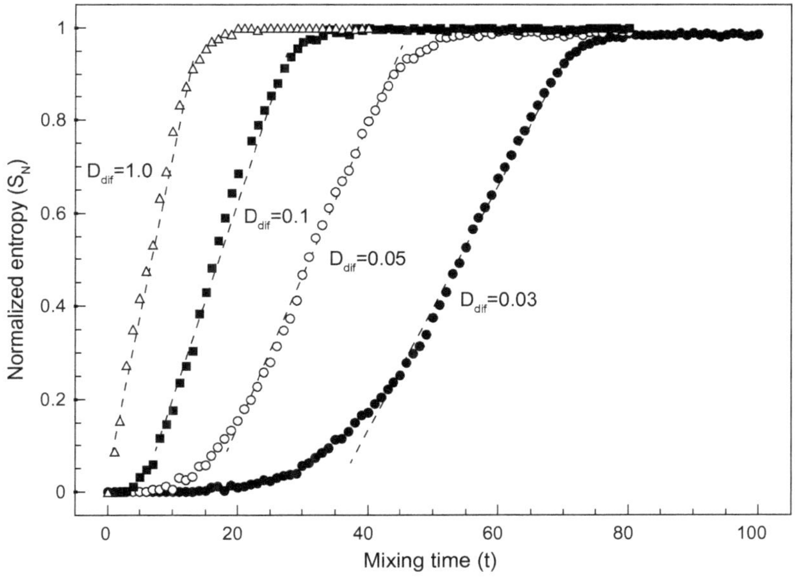

Fig. 4.13 Variation of Shannon entropy against mixing time (t). The fitting of the linear portions of the plots is represented by the dashed lines

where $p_i(t)$ is the probability that hybrid data points fall into cell c_i at time t. $p_i(t)$ is calculated for each cell as the ratio between the number of points falling in the cell (N_c) and the total number of points (N_{tot}). Thus, $S(t)$ is estimated by summing the values of $p_i(t) \log p_i(t)$ calculated for all cells M. The least efficient mixing configuration (i.e. before the beginning of the mixing process) is represented by $S = 0$ (i.e. no data point has the hybrid composition). The most efficient mixing configuration (i.e. all cells contain the same number of data point having the hybrid composition) is, on the contrary, represented by $S_{\max}$. Normalization of S_N to the maximum value ($S_{\max}$) allows us to compare the evolution of S for different elements. S is a quantity that increases with mixing time (Baranger et al. 2002).

Given the initial proportions of white and black colored magma (70% and 30%, respectively; Fig. 4.10), the hybrid composition for the numerical simulation corresponds to a value of 0.7. The bar-code representation of the hybrid data point distribution for the numerical simulations is reported on the top of the compositional series shown in Figs. 4.11 and 4.12. The graphs show that, as mixing time increases, a progressively larger amount of hybrid samples are formed. This produced increasingly dense bar-code plots (Fig. 4.11). Furthermore, chemical elements with different diffusivities show different bar-code plots at the same time and in the same mixing system. In detail, the faster the element is, the larger the amount of hybrid samples (Fig. 4.12).

Figure 4.13 shows the variation of (S_N) as a function of mixing time (n). The graph shows that S_N and mixing time are positively correlated. For long mixing times S_N saturates to constant values, corresponding to an homogeneous system. This is typical for all elements, with the major difference that the slope (dS/dt) of the rectilinear segments of the curves increases with increasing D_{dif}. The slope coefficient (i.e. dS/dt) for each element, estimated by linear fitting, highlights that dS/dt increases from ca. 0.021 to 0.078 as the diffusion coefficient increases.

The plot of Fig. 4.14 displays the variation of Shannon entropy (dS/dt) against RCV. A linear correlation between these two variables is evident: values of dS/dt increase as the mobility of chemical elements (RCV) increases.

These results are potentially of great importance to enhance our ability to study the complexity of magma mixing processes and, in particular,

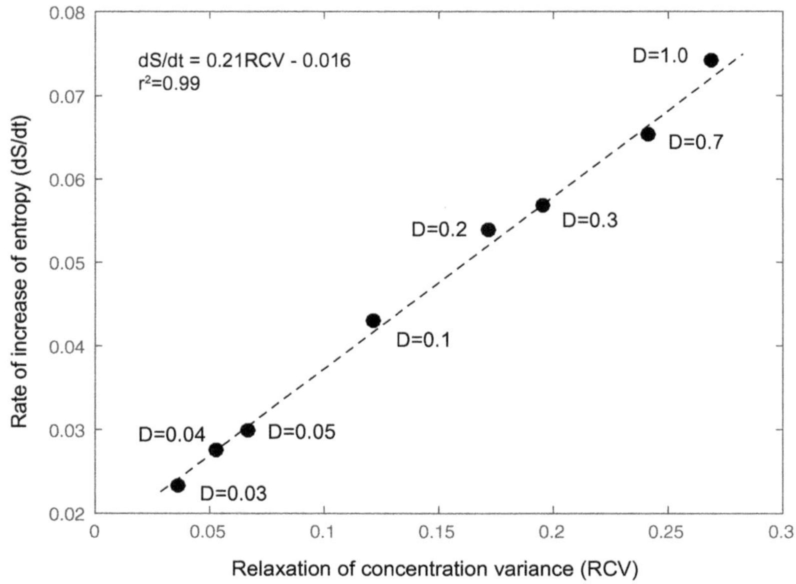

Fig. 4.14 Relaxation of Concentration Variance (RCV) against the rate of increases of entropy (dS/dt). Best-fit line and linear equation are also displayed

to reconstruct the original compositions of interacting melts as well as identifying the hybrid composition that the system would eventually attain.

4.2.5 Compositional Histograms and Hybrid Composition

We now discuss a method that can be helpful to identify the initial proportions of liquids starting from the analysis of concentration patterns generated by the mixing process. This aspect is particularly relevant in the study of natural cases because we generally deal with snapshots of the mixing process where the advection/diffusion dynamic couple might have acted with different efficiency. Figure 4.15a shows the variation of frequency histograms for the system reported in Fig. 4.10. The graph shows that as the mixing time (n) increases, histograms evolve to a bell-shape and all compositions between black (mafic) and white (felsic) magmas are present in different amounts, with intermediate compositions having the highest frequency. The graph of Fig. 4.15b shows that the variation in the initial percentages of magmas in the simulations corresponds to a shift of the maximum frequency of the histograms towardsdifferent compositions. A detailed observation of Fig. 4.15b indicates that this maximum corresponds to the hybrid composition that the system would eventually attain if the mixing was completed. This can be tested using the linear two end-member mixing equation:

$$C_h = C_f x + C_m (1 - x) \quad (4.9)$$

where C_h, C_f and C_m are the composition of the hybrid, felsic and mafic magma, respectively, and x is the proportion of felsic magma in the mixture. For instance, the theoretical hybrid for a system constituted by 70% of felsic (white colored) magma is ca. 0.7 [i.e. $C_h = 1.0 \cdot 0.70 + 0 \cdot (1 - 0.70)$], which corresponds exactly to the maximum value of the histogram (Fig. 4.15). The reader can verify the other values of C_h for the histograms in Fig. 4.15b. Therefore, the maximum frequency of compositional histograms corresponds to the theoretical hybrid composition of the mixing system.

These results will be utilized in the next chapter to estimate the initial percentages of end-member magmas in natural samples. Indeed, knowing the composition of the hybrid magma, i.e. the maximum frequency of the histograms, and the compositions of the end-member mafic and felsic mag-

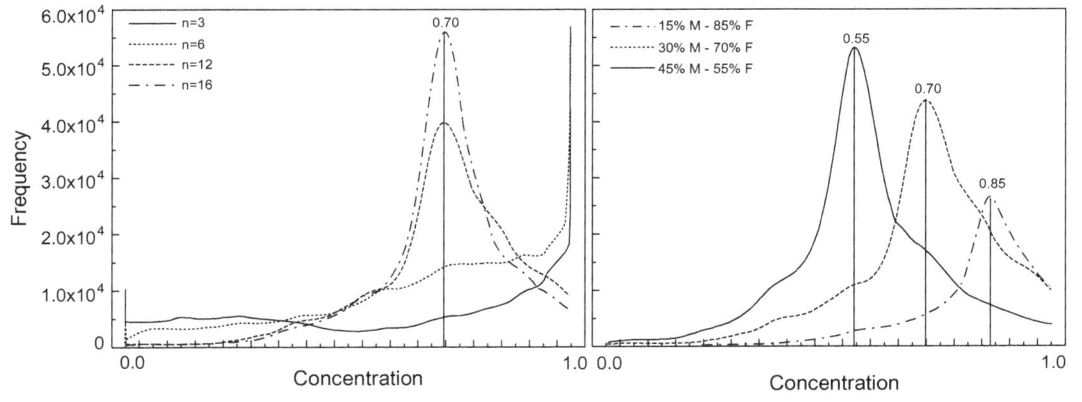

Fig. 4.15 Frequency histograms of concentration of the mixing systems resulting from the numerical simulation of magma mixing. **a** Relative proportions of black colored mafic and white colored felsic magma are 30% and 70%, respectively. **b** Frequency histograms for mixing systems with different initial amounts of mafic and felsic magma at the same number of iterations ($n = 12$)

mas, the percentages of magmas participating to the mixing process can be estimated.

4.3 Three-Dimensional Modeling

4.3.1 Stretching and Folding (Advection)

The numerical modeling scheme that we consider for advection in three dimensions is the "ABC map" (e.g. Cartwright et al. 1994; Galluccio and Vulpiani 1994; Haller 2001). This flow scheme reproduces mixing by stretching and folding (e.g. Perugini et al. 2002, 2007).

The system is characterized by the presence of two dynamic regions: (i) Coherent Regions (CR; i.e. regular islands) and (ii) Active Mixing Regions (AMR; i.e. chaotic regions), as for the sine flow map, but in $3D$.

The ABC map which we consider here is defined as:

$$\begin{aligned} x_{n+1} &= x_n + A\sin(z_n) + C\cos(y_n) \bmod [2\pi] \\ y_{n+1} &= y_n + B\sin(x_{n+1}) + A\cos(z_n) \bmod [2\pi] \\ z_{n+1} &= z_n + C\sin(y_{n+1}) + A\cos(x_{n+1}) \bmod [2\pi] \end{aligned} \tag{4.10}$$

where A, B, and C are the parameters of the map, and x_n, y_n, z_n and x_{n+1}, y_{n+1}, z_{n+1} are the coordinates of each fluid particle at iteration n and $n+1$, respectively. It forms a periodic three-dimensional chaotic system whose domain is $0 \le (x, y, x) \le 2\pi$.

Perugini et al. (2007) have suggested that values of $A = 1.0$, $B = 0.82$ and $C = 0.58$ appear reasonable to reproduce, as a first approximation, magma mixing systems constituted by both AMR and CR, as observed in natural samples.

As discussed for the sine flow map, the rate of stretching (λ) is a crucial factor for mixing. Figure 4.16a shows the stretching field in the ABC map and reveals the presence of CR (yellow colour) and AMR (blue colour). CR in $3D$ are constituted by tubes (Fig. 4.16b) in the $3D$ volume. These are surrounded by high stretching regions (AMR). The influence of CR and AMR on the deformation of two spherical blobs of magma (Fig. 4.17a) is displayed in Fig. 4.17b–d. The blob in the CR undergoes little deformation whereas the blob in the AMR undergoes vigorous stretching and folding dynamics.

The morphology attained by the initial spherical blob in the CR during the advection process is shown in Fig. 4.18. Petrologists and volcanologists involved in the study of magma mixing processes can probably recognize the similarity between the shapes shown in the figure and those observed on natural outcrops.

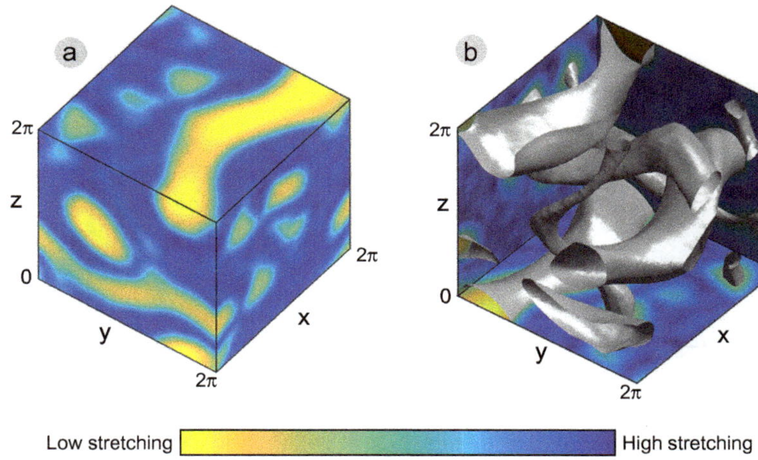

Fig. 4.16 **a** Stretching field (λ) calculated for the ABC map; **b** tubes of low stretching (CR) within the system domain. To improve visualization, planes of stretching at $x = 0$, $y = 0$, and $z = 0$ are also shown

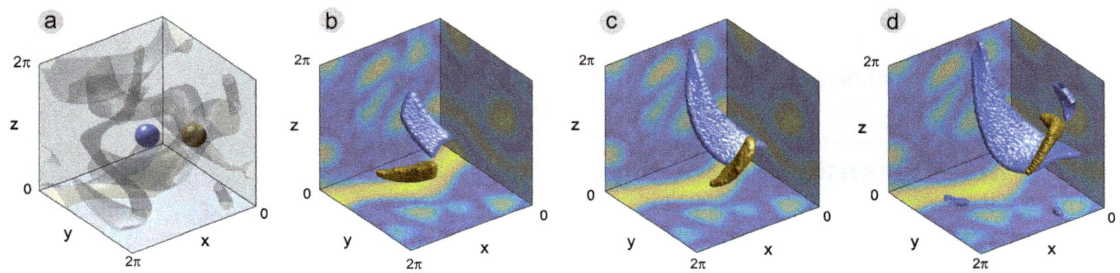

Fig. 4.17 **a** Starting configuration of the mixing system in which a blue and a brown spherical portion of magma are positioned initially in the AMR and CR, respectively: the low stretching tubes are also shown; **b–d** evolution of the mixing system after two (**b**), four (**c**), and seven (**d**) iterations. Planes of the stretching field at $x = 0$, $y = 0$, and $z = 0$ are also reported

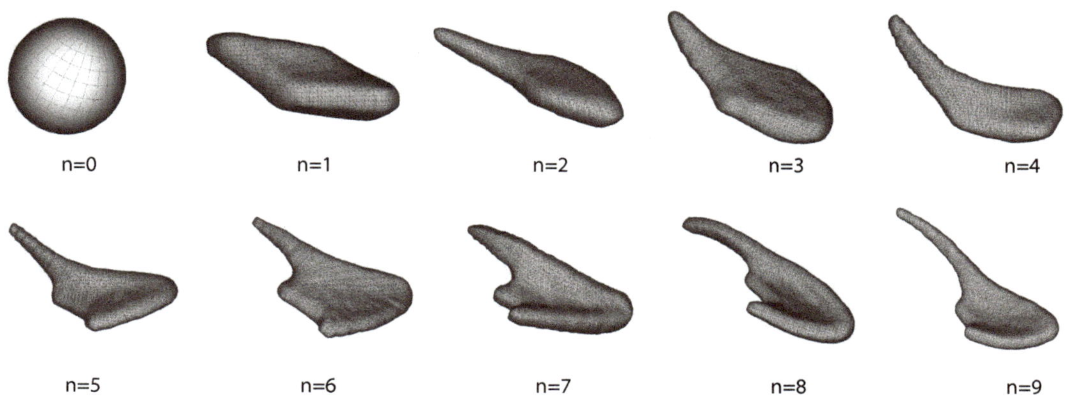

Fig. 4.18 Shape variation of the blob originally positioned into the CR with the progression in time of the mixing process

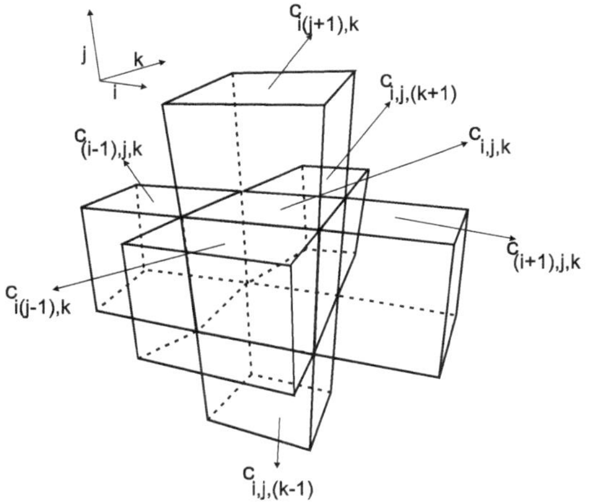

Fig. 4.19 Portion of the $3D$ grid used in the discretization of the ABC map domain for the numerical simulation of chemical diffusion

4.3.2 Advection and Diffusion

To simulate the chemical diffusion process, the model of chaotic advection is coupled with the finite difference algorithmic approach of diffusion reported in Pierrehumbert (1994), but in 3D (Perugini et al. 2007; Fig. 4.19). We define a continuous concentration field $c(x, y, z)$, and spatially discretized it to a regular $3D$ grid $c_{i,j,k} = c(x_i, y_j, z_k)$. The map is first used to rearrange $c(x, y, z)$ through mapping (x, y, z) and re-interpolating to the grid. The diffusive step is implemented as:

$$c_{i,j,k} = (1 - D_{\text{dif}})\, c_{i,j,k} + \Delta c_{i,j,k} \qquad (4.11)$$

where $\Delta c_{i,j,k}$ is

$$\Delta c_{i,j,k} = \frac{1}{6} \cdot D_{\text{dif}} \cdot \big[c_{(i+1)jk} + c_{(i-1)jk} + c_{i(j+1)k} + c_{i(j-1)k} + c_{ij(k+1)} + c_{ij(k-1)}\big] \qquad (4.12)$$

As in the case of the $2D$ sine flow map, D_{dif} is the diffusion coefficient varying between zero and unity and i, j and k are the indices of the lattice. During this step, $c(i, j, k) = c(x_i, y_j, z_k)$ is updated to new values according to the diffusion that each cell underwent as calculated using the above equation. After several iterations of the map, coupled with the diffusion step, the concentration field $c(x, y, z)$ will have changed. The value of D_{dif} used in the simulations is 0.9.

The $3D$ simulation is performed considering the initial configuration of Fig. 4.20a. Here spherical portions of mafic magma (30% in volume) with an initial concentration equal to zero are distributed in a felsic magma (70% in volume) with an initial concentration equal to unity.

As the advection/diffusion simulation progresses in time, intermediate concentrations are generated. The evolution of the system for different iterations is shown in Fig. 4.20b–d. Here, to visualize the change in composition, the $3D$ volume has been sliced by a $2D$ plane. As the mixing process proceeds in time the homogeneity of the system increases. This is shown in the compositional histograms of Fig. 4.21. If no chemical diffusion between the two magmas occurs, the compositional histogram will be constituted by two populations, corresponding to the initial compositions of the mafic and felsic magma. In the presence of chemical diffusion, on the contrary, the compositional histogram will evolve to a normal distribution with increasing mixing time. In this case the histogram will be populated by all compositions between those of the initial mafic and felsic magma. As shown previously, the modal value of the histogram coincides with the hybrid composition (Perugini et al. 2004).

As we have seen in this chapter, numerical models can aid in conceiving and visualizing the spatial and temporal variability of magma mixing processes. We have also shown that there is not significant difference between two-dimensional and three-dimensional simulations. The use of one or the other depends on the type of application. Numerical models can provide new tools for the quantification of mixing processes, such as the decay of concentration variance and the use of histograms to derive the hybrid composition of the mixing system. In the next chapters we will refer frequently to the results from the numerical simulations presented above in the attempt to provide an explanation for both the structures and compositional patterns observed in natural rock sam-

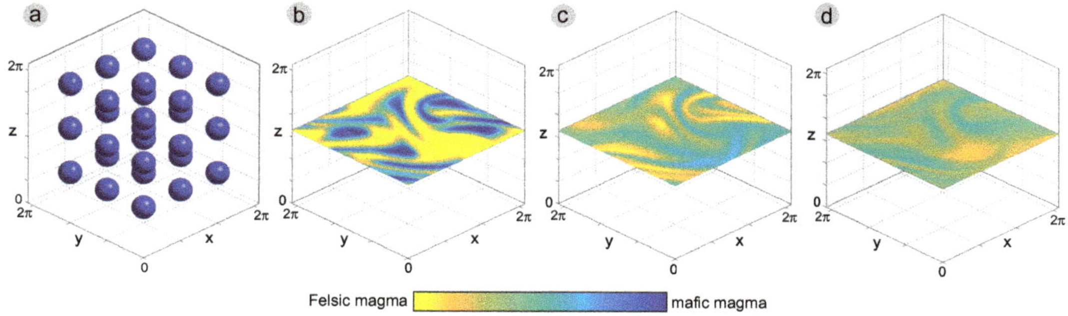

Fig. 4.20 **a** Initial configuration used in the advection-diffusion numerical model; **b–d** $2D$ slices placed at $z = \pi$ showing the change of concentration in the $3D$ system as the mixing time increases

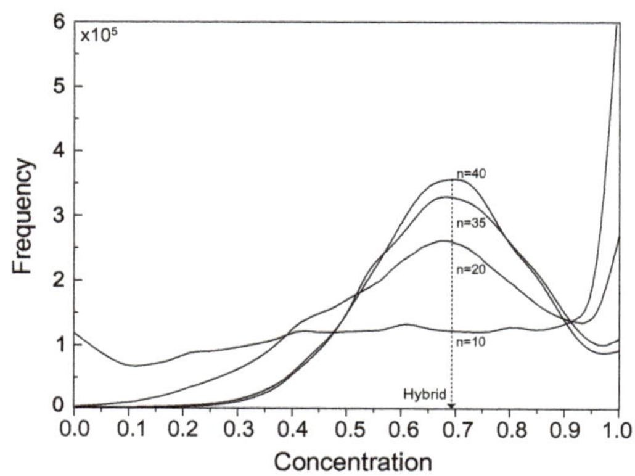

Fig. 4.21 Frequency histograms for the whole $3D$ system showing the convergence of the concentration spectrum to a bell-shape as the number of iterations increases. The maximum of the curve corresponds to the composition of the hybrid that the mixing system would eventually attain at $n = \infty$

ples. Also, we will try to highlight advantages and disadvantages in using these models during their application to natural case studies.

References

Aref H, El-Naschie MS (1995) Chaos applied to fluid mixing. Pergamon Press, Reprinted from Chaos, Solitions and Fractals 4(6)

Baranger M, Latora V, Rapisarda A (2002) Time evolution of thermodinamic entropy for conservative and dissipative chaotic maps. Chaos Solitions Fractals 13:471–478

Camesasca M, Kaufman M, Manas-Zloczower I (2006) Quantifying fluid mixing with the shannon entropy. Macromol Theory Simul 15(8):595–607. ISSN: 1022-1344. https://doi.org/10.1002/mats.200600037

Cartwright JHE, Feingold M, Piro O (1994) Passive scalars and three-dimensional Liouvillian maps. Physica D 76:22–33

Clifford MJ, Cox SM, Roberts EPL (1998) Lamellar modelling of reaction, diffusion and mixing in a two-dimensional flow (English). Chem Eng J 71(1):49–56

Clifford MJ (1999) A Gaussian model for reaction and diffusion in a lamellar structure (English). Chem Eng Sci 54(3):303–310

Collet P, Eckman JP (1980) Iterated maps on the interval as dynamical systems. In: Ruelle AJ (ed.). Birkhauser

Crank J (1975) The mathematics of diffusion. Clarendon Oxford

Galluccio S, Vulpiani A (1994) Stretching of material lines and surfaces in systems with Lagrangian Chaos (English). Physica A 212(1–2):75–98

Haller G (2001) Distinguished material surfaces and coherent structures in threedimensional fluid flows (English). Physica D 149(4):248–277

Jain AK (1989) Fundamentals of digital image processing. Prentice-Hall, Hoboken

Liu M et al (1994) Structure of the stretching field in chaotic cavity flows (English). AIChE J 40(8):1273–1286

Liu WJ, Haller G (2004) Strange eigenmodes and decay of variance in the mixing of diffusive tracers (English). Physica D-Nonlinear Phenomena 188(1–2):1–39

Liu M, Muzzio FJ, Peskin RL (1994) Quantification of mixing in aperiodic chaotic flows (English). Chaos Solitons & Fractals 4(6):869–893

Mathew G et al (2007) Optimal control of mixing in Stokes fluid flows (English). J Fluid Mech 580:261–281

Ott E, Antonsen TM (1989) Chaotic fluid convection and the fractal nature of passive scalar gradients. Phys Rev Lett 61:2839–2842

Perugini D et al (2004) Kinematic significance of morphological structures generated by mixing of magmas: a case study from Salina Island (southern Italy) (English). Earth Plan Sci Lett 222(3–4):1051–1066. ISSN: 0012821X. https://doi.org/10.1016/j.epsl.2004.03.038

Perugini D, Petrelli M, Poli G (2007) Influence of landscape morphology and vegetation cover on the sampling of mixed plutonic bodies (English). Mineral

Petrol 90(1–2):1–17. ISSN: 09300708. https://doi.org/10.1007/s00710-006-0173-1

Perugini D, Poli G, Gatta GDD (2002) Analysis and simulation of magma mixing processes in 3D (English). Lithos 65(3–4):313–330. ISSN: 00244937. https://doi.org/10.1016/S0024-4937(02)00198-6

Perugini D, Poli G, Mazzuoli R (2003) Chaotic advection, fractals and diffusion during mixing of magmas: Evidence from lava flows. J Volcanol Geother Res 124(3–4):255–279. ISSN: 03770273. https://doi.org/10.1016/S0377-0273(03)00098-2

Pierrehumbert RT (1994) Tracer microstructure in the large-Eddy dominated regime (English). Chaos Solitons & Fractals 4(6):1091–1110

Rothstein D, Henry E, Gollub JP (1999) Persistent patterns in transient chaotic fluid mixing (English). Nature 401(6755):770–772

Sepulveda MA, Badii R, Pollak E (1989) Spectral analysis of conservative dynamical systems (Eng). Phys Rev Lett 63(12):1226–1229

Shannon CE (1948) A mathematical theory of communication. Bell Syst Techn J 27:379-423. ISSn: 07246811

5 Experiments

It doesn't matter how beautiful your theory is ... If it doesn't agree with experiment, it's wrong.

Richard Feynman

Abstract

Physical modelling allows for the study of new processes and the evaluation of variables in complex flows which cannot be accessed by theory and numerical models. Physical modelling also allows us to validate theoretical results, analyse different details of a phenomenon and test a variety of environmental and extreme conditions. This is particularly relevant in the study of magmatic processes. This chapter reports on the applicability of experiments in the context of magma mixing. We review the state of the art and illustrate in detail one of the most advanced experimental setups, the chaotic magma mixing apparatus, available today to face the complex challenge to understand magma mixing processes and their impact upon our knowledge of magmatic systems.

5.1 The Meaning of Experiments

The technological transformation that occurred in the last decades allowed us to develop new sophisticated equipment and to improve considerably experimental techniques. Thanks to this, physical modelling allows for the study of new processes and the evaluation of variables in complex flows which cannot be accessed by theory and numerical models. Physical modelling also allows us to validate theoretical results, analyse different details of a phenomenon and test a variety of environmental and extreme conditions. This is particularly relevant in the study of magmatic systems.

Experiments can be helpful in combination with numerical simulations to define hybrid approaches to benefit from the advantages of each model. In a hybrid model, we can define numerical-physical or physical-numerical perspectives. For example, we can use a numerical simulation to produce inputs to the experiment. This, in turn can provide information to refine the same numerical model or for the interpretation of a natural process.

The outputs from numerical models and experiments are essential for understanding the behaviour of magmatic systems, including magma mixing and, therefore, their integration is critically important.

In principle, in order to draw the most complete picture about a natural process, an experimental program needs a large number of experiments; however, experiments, in the context of experimental petrology and volcanology, are generally time-consuming and very expensive. From

D. Perugini, *The Mixing of Magmas*, Advances in Volcanology,
https://doi.org/10.1007/978-3-030-81811-1_5

this point of view, numerical modelling can play an important role in balancing these drawbacks, as it can aid to interpret the experimental results and, thus, reduce the number of experiments to be performed.

In this chapter we discuss the applicability of experiments in the context of magma mixing. We briefly review the state of the art and illustrate in detail one of the most advanced experimental setups (the chaotic magma mixing apparatus) available today to face the complex challenge to understand magma mixing processes and their impact upon our knowledge of magmatic systems.

5.2 Experimental Mixing of Magmas

Several attempts have been made to simulate experimentally the process of magma mixing. Different experimental setups have been used and this allowed to obtain different bits of information. Here we will mainly focus on experiments performed at high-temperature and high-pressure using silicate melts, with compositional and rheological properties comparable to those of natural magmas.

Before the beginning of the new century the only attempts to reproduce experimentally magma mixing were made by Kouchi and Sunagawa (1983) and Kouchi and Sunagawa (1985). They used natural basaltic and dacitic compositions as and-member magmas.

Two rods with different composition were melted at their relative ends at approximately 1250 °C, connected and rotated in opposite directions to trigger magma mixing by forced convection. Experiments showed that the two magmas could be easily mixed to form both banded dacite and homogeneous andesite, in timescales of the orders of a few hours. The presence of phenocrysts in the magmas modified the flow patterns making the mixing process more efficient. These experiments indicated that mixing at larger length scales progressed basically through mechanical mixing. As mixing time increased, and the length scales between the two magmas decreased, chemical diffusion dominated the scene leading to hybrid compositions in short time.

These very first magma mixing experiments bring us a clear message: dynamics conditions (in this case forced convection) are a prerequisite to disperse one magma into the other from the macro- to micro-scale. Further, the generation of small-scale lamellar patterns between the interacting magmas is essential to trigger efficient chemical diffusion and to generate hybrid compositions.

The 2000s have seen a growing interest in magma mixing experiments. De Campos et al. (2004) performed mixing experiments between trachytic and phonolitic trachytic compositions. To enhance mixing, the initial magmas contained partially dissolved crystals. The two starting compositions were loaded as cylinders and convection was forced via rotation of a spindle in a Taylor-Couette-like experimental setup. The experiments were performed at 1300 °C. They recognized the development of two separate convection cells in the experimental samples: a lower one with a primary flow parallel to the bottom and side walls of the crucible, and an upper one, with a flow approximately at right angle to that of the lower cell. Mixing was essentially driven primarily by forced convection combined with the effect of local compositional gradients (diffusion) leading to a density distribution similar to a double-diffusive system. Again, advection-diffusion dynamics were observed to be the main forces promoting mixing.

Zimanowski et al. (2004) conducted experiments using basaltic and rhyolitic compositions, inducing the mixing process by remelting natural volcanic rocks in stratification configuration using a high-temperature rotational viscometer. The idea was mainly to test experimentally the observations in natural rocks of drop-like domains of less evolved magma in more evolved magma and to evaluate the contribution of mechanical mixing processes (called by authors liquid immiscibility) in their formation. Experiments were carried out with the aim to validate model calculations for shear rate dependent domain sizes. The results confirmed the model and authors argued that, in the context of volcanism, the key process that is able to change the state of a magma reservoir on a critical time scale is hydrodynamic magma mixing.

Laumonier et al. (2015) performed deformation and mixing experiments using water-saturated magmas with basaltic and haplotonalitic composition at 300 MPa, in the temperature range 600–1020 °C. Before mixing, the two magmas were allowed to crystallize, yielding magmas with crystal contents in the range 31–53 wt.%. They found that mixing/mingling textures started to appear in the experiments at temperatures of ca. 950 °C. In the temperature range 950–985 °C, a few mixing and mingling textures occurred and both end-members essentially retained their physical integrity. A dramatic jump in mingling efficiency was observed at, or above, 1000 °C, corresponding to a crystal fraction of 45 vol.%. Textures included entrapment of mafic crystals into the felsic magma, mafic–felsic banding, enclave formation, and diffusion-induced interfaces. In the most strained parcels of interacting magmas, complex mixing/mingling textures were produced, similar to those observed in volcanic and plutonic rocks. The experiments showed that mixing between hydrous felsic and mafic magmas is likely to take place efficiently at around 1000 °C. These authors also argued that magma mixing might trigger volcanic eruptions driving only small (ca. 15 °C) temperature fluctuation in the sub-volcanic reservoir.

The same research group (Laumonier et al. 2014) carried out further mixing experiments with the aim of developing physical simulations relevant to magmatic reservoirs in the upper crust. Torsion experiments on felsic-mafic-stacked layers, using an internally heated deformation apparatus, were performed. The torsion experiments involved fully molten and crystal-bearing felsic end-members and a crystal-bearing mafic magma. They found that hybridization during single-intrusive events requires injection of high proportions of the intruding basaltic magma during short periods, producing magmas with 55–58wt.% SiO_2. Authors also suggested that high strain rates and gas-rich conditions are likely to produce more felsic hybrids. They argued that the incremental formation of crustal magma chambers can limit the generation of hybrids to the very last stages of pluton formation and to small volumes. On the other hand, large-scale mixing appears to be more efficient at lower crustal conditions, but requires higher proportions of mafic melt, producing more mafic hybrids than in shallow reservoirs.

In the second half of 2010 a massive experimental work on magma mixing started. This was developed in the framework of a collaborative effort between the Department of Physics and Geology of the University of Perugia (Italy) and the Department of Earth and Environmental Sciences, Ludwig-Maximilians-University (LMU) of Munich (Germany). At LMU we carried out mixing experiments in several configurations, including rotational dynamics induced by the Taylor-Couette scheme, injection of mafic magmas into felsic magmas using a centrifuge apparatus, and the development of a prototype of the so-called journal bearing flow. The latter was further developed at the Department of Physics and Geology of the University of Perugia where, at present, it represents the most advanced equipment to perform chaotic mixing of magmas. In the following we will describe in details the experiments performed with the centrifuge, deferring a detailed description of the chaotic magma mixing apparatus (COMMA) to the next section.

5.2.1 A Centrifuge to Perform Magma Mixing Experiments

The experimental apparatus consists of a standard laboratory centrifuge that was modified to hold a high-temperature (1200 °C) furnace (Dorfman et al. 1996). Temperature was controlled by three independent Pt-$Pt_{90}Rh_{10}$ (type S) thermocouples at the top, bottom and the middle of the sample crucible. In this configuration a small thermal-gradient (<1 °C over about 15–20 mm) was obtained.

A sealed Pt-capsules hosted cylinders of end-members, in a buoyancy unstable geometry (least dense melt, hereafter felsic, in the outer part and most dense, hereafter mafic, in the inner part; Fig. 5.1). Experiments were performed at 1200 °C with an acceleration of 10^3g ($g = 9.81\,m/s^2$).

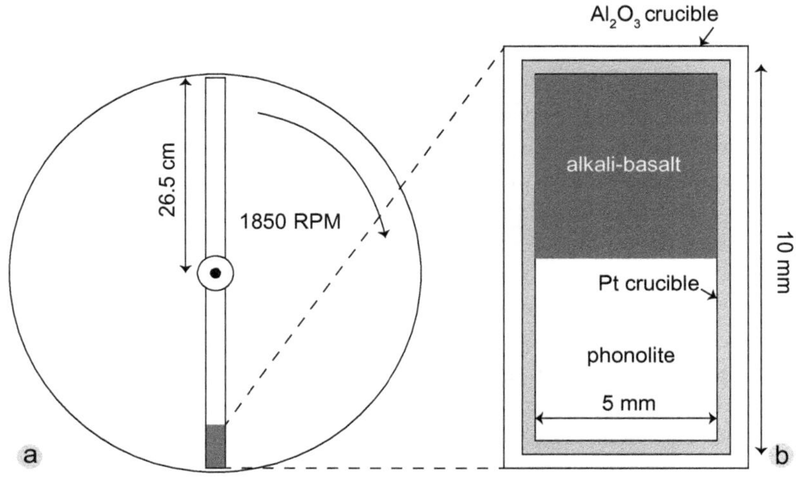

Fig. 5.1 Schematic representation of the centrifuge apparatus used to perform magma mixing experiments (**a**) and sketch of the experimental capsule containing the mafic and felsic magma (**b**)

Three experiments with identical starting conditions were performed with different run duration (5, 20 and 120 min).

After each experiment, the sample was quenched in air and the resulting glasses were cut in half parallel to their cylindrical axis and analyzed for their major and trace elements by electron microprobe and Laser Ablation ICP-MS (Perugini et al. 2013).

The geometric configuration adopted in the experiments was designed to inject the mafic magma in the felsic one. By forcing the mafic melt towards the opposite side of the capsule, the felsic melt was driven to the other side. During rotation, volumes of the mafic melt were incorporated into the felsic one and were pulled toward the external walls of the crucible. This way mixing mainly occurred at the interface.

The Reynolds number (Re) of these experiments was of the order of 10^{-10}, estimated as

$$\mathrm{Re} = \frac{F_c d \rho}{35 \pi l \mu^2} \tag{5.1}$$

(e.g. Yuan and Zheng (2011)) where $F_c = \rho V \omega^2 R_c$ is the centrifugal force, ρ is density, V is the capsule volume, ω is the angular velocity of the centrifuge, R_c is the distance between the center of rotation and the experimental capsule, d and l are the capsule diameter and length, and μ is the viscosity.

Experiments were performed using natural compositions from the Phlegrean Fields (Italy) volcanic area (Arienzo et al. 2009). In particular, the end-member melts derive from Agnano Monte-Spina B1-B2 phonolitic tuff (de Vita et al. 1999) and the alkali-basaltic glass matrix from the Minopoli eruption (Di Vito et al. 1999). Measured viscosities and calculated densities of the basaltic and the phonolitic melt at 1200 °C (i.e. the temperature used during the experiments) are $2 \cdot 10^2$ and $2 \cdot 10^4$ Pa s, and 2.82 and 2.50 g/cm^3, respectively.

Experimental samples after run duration of 5, 20, and 120 min experiments are shown in Fig. 5.2. The sample quenched after 5 min depicts the mafic, denser melt, moving to the side opposite its initial location due to the centrifugal force (Fig. 5.2a). Filaments of the mafic melt were captured in the felsic one. At the interface, the felsic melt was partly incorporated into the mafic one. The evolution of the mixing system towards a larger efficiency is clearly shown in the 20 min sample (Fig. 5.2b). In detail, the filaments appear collapsed at the melt interface. Here, the development of vortexes entraining the two melts occurred. The experiment performed with the longest time (120 min) shows that the experimental sample is divided into two parts (Fig. 5.2c): the mafic melt at the bottom and the felsic one at the top of the crucible. Ghosts of filament-like patterns are still

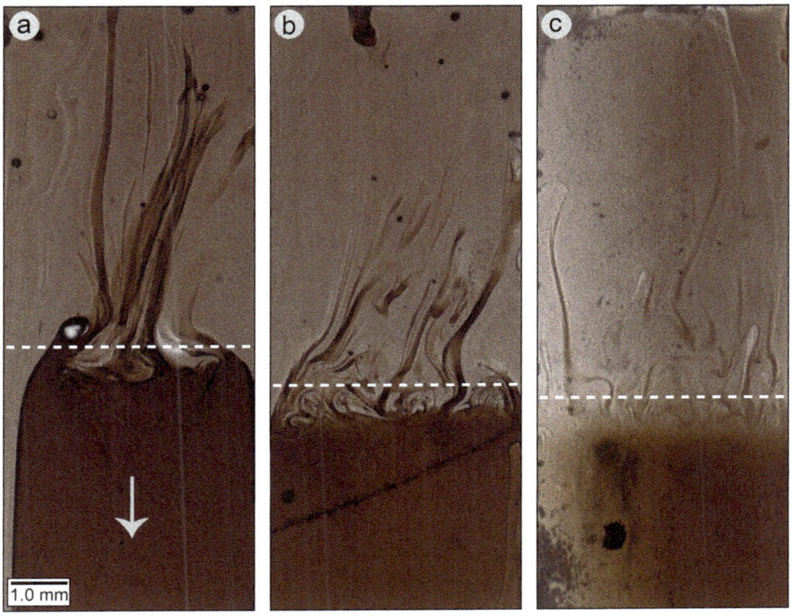

Fig. 5.2 Optical images of experimental samples at mixing time $t = 5$ min (**a**), $t = 20$ min (**b**) and $t = 120$ min (**c**). The white arrow in **a** points to displacement of the mafic melt during the experiment. Dashed lines indicate the path followed by micro-analysis. Modified from Perugini et al. (2013)

observable in the middle part of the experimental sample.

The compositional variation produced by the mixing process was quantified along transects crossing the interface where most of mixing occurred (white dashed lines; Fig. 5.2).

Before discussing any aspect related to the compositional variations in the experimental samples it is important to decipher the relative proportions of the two melts interacting in the investigated sample segment. As we have discussed in Chap. 4, the chemical variability produced during mixing can be used to infer the initial proportions of melts by knowing the initial concentration of chemical elements in the two starting melts and the theoretical hybrid concentration (e.g. Perugini et al. 2004; Perugini et al. 2008). This method is based on the study of compositional histograms. Figure 5.3a displays a representative compositional histogram. As discussed in Chap. 4, compositional histograms for a mixing system should evolve in time from the original end-member compositions to a bell-shape whose maximum corresponds to the hybrid composition.

The plot of Fig. 5.3a agrees with these considerations. As the mixing process evolves in time, the compositional variability narrows and the histogram becomes a bell-shape curve for the longest mixing time (i.e. 120 min). Given that the maximum of the curve corresponds to the hybrid composition at which the system will tend in the long time (e.g. Perugini et al. (2004)), we can use the two end-member mixing equation to infer the initial proportions of melts that mixed along the transects (see Chap. 4 for details). Before doing this, however, the hybrid composition must be estimated. This can be done using the concentration histograms for the 120 min experiment (i.e. the one with the longest mixing time) and fitting by a Gaussian function of the form:

$$y = A \exp\left[\frac{-(x - x_c)^2}{2w^2}\right] \tag{5.2}$$

where x is the concentration of a given element, y is concentration frequency, and A, w, x_c are the height, width and center of the Gaussian function, respectively. x_c is the parameter of interest (i.e. the hybrid concentration, $x_c = C_h$, see Eq. 4.9 in Chap. 4). Figure 5.3a shows the fitting procedure for the compositional histogram of Ba and the determination of the hybrid concentration (C_h). This procedure was repeated for all analyzed major and trace elements. The plot in Fig. 5.3b shows that the

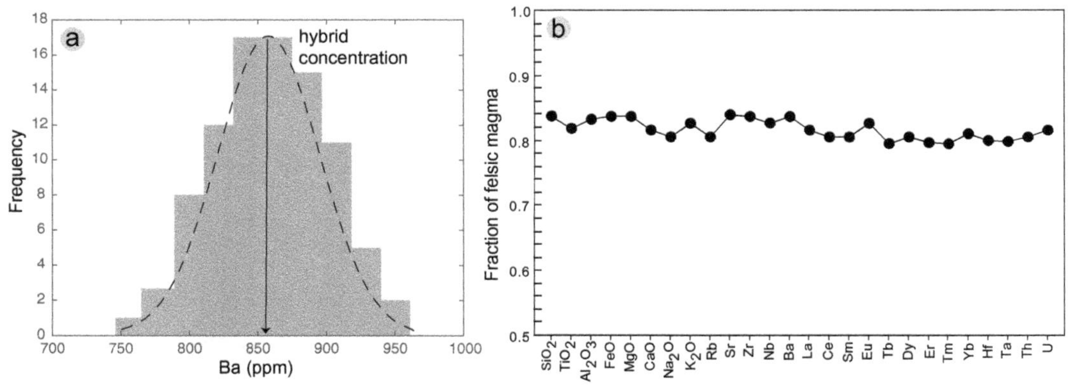

Fig. 5.3 **a** Fitting of the compositional histogram of Ba ($t = 120$ min experiment) using the Gaussian function given by Eq. 5.2 and determination of the hybrid concentration for this element; **b** fraction of felsic melt for each chemical element, estimated using Eq. 4.9

fraction of the felsic melt can be estimated in the range of 0.78–0.84 and, consequently the mafic melt fraction is in the range 0.22–0.16. Therefore, the compositional variability studied along the interface is due to the mixing of basalt and phonolitic melts in the proportion of ca. 20:80. It is interesting to note how all analyzed chemical elements converge to a narrow range of felsic melt fractions indicating that the compositional histogram method can provide reliable and statistically robust results (Fig. 5.3b).

The variation of representative major and trace elements along the three transects is shown in Fig. 5.4. At short mixing times (i.e. 5 min), element variability oscillates across the transect according to the occurrence of filaments of the two melts. The transition between the different filaments increases (or decreases) continuously. This is an indication that chemical exchanges operated between the two melts.

After 20 min (Fig. 5.4) no trace of the starting mafic melt can be detected along the transects. Rather, the compositional field is shifted towards the hybrid composition. The amplitude of compositional fluctuations is strongly reduced and this is matched by an increase of frequencies (i.e. number of filaments). These effects are related to the evolution in time of the mixing process: repeated stretching and folding produced an increase of the number of filaments. The increase of interfacial area generated by these dynamics enhanced chemical exchanges, progressively shifting the system composition towards the hybrid. After 120 min (Fig. 5.4) the compositional variability along the interface between the two melts is strongly reduced and the compositional field has collapsed towards the hybrid. Compositional gradients appear smoothed out by the efficient advection and diffusion and the system appears rather homogeneous.

Figure 5.5 displays binary inter-elemental plots of some representative major and trace elements for the three experiments. The evolution of the compositional field is marked by a continuous narrowing of the compositional variability in time, with data points progressively clustering towards the hybrid composition. In these plots major elements mostly describe linear variations, with the exception of Na_2O (Fig. 5.5) that defines sigmoidal patterns. As for trace elements, some pairs show linear patterns (Fig. 5.5), while others define more complex variations with variable scattering of data points (Fig. 5.5). An explanation of this variability in inter-elemental correlations will be provided in the next chapters. Here it suffices to say that the patterns in Fig. 5.5 are most probably related to the different mobility of chemical elements in the magmatic system during the development of the mixing process.

As discussed in Chap. 4, a possible method to quantify the chemical element mobility is to estimate the Relaxation of Concentration Variance

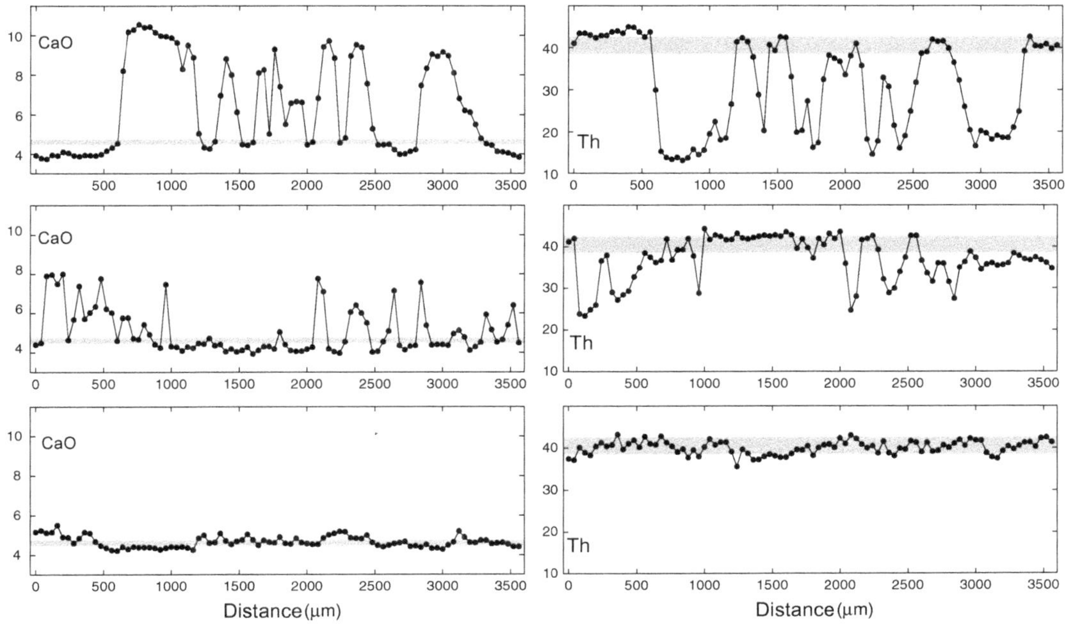

Fig. 5.4 Compositional variation of two representative chemical elements across the mafic-felsic interface (Fig. 5.2) after 5, 20, and 120 min (mixing time increases from top to bottom). The concentration of the theoretical hybrid composition is marked as a gray strip to include analytical uncertainties

(RCV). We recall that this parameter is the rate of decay of concentration variance during mixing (parameter R in Eq. 4.6). Concentration variance (σ_n^2) was calculated for all analyzed chemical elements at different mixing times (i.e. 5, 20, and 120 min). Results show that for all elements concentration variance decreases exponentially in time, as it should be expected because of the increasing mixing efficiency (Fig. 5.6). This is in agreement with the results obtained from numerical simulations reported in Chap. 4. In addition, the rate of decay of concentration variance (σ_n^2) is different for the different elements. For example, considering major elements, the rate of decay increases from SiO_2 to Na_2O. Analogous considerations can be made for the trace elements, where the rate of decay increases from Rb to Yb.

The graph of Fig. 5.7 shows estimated values of RCV for all analyzed chemical elements plotted against their ionic radius. Na displays the largest mobility followed by Al. Moving to lower RCV we observe a zone in which RCV is in the range 1.0–0.08 including the other major elements (i.e. Mg, Ca, K, Si, Ti and Fe) and trace elements such as (Nb, Ta, Zr, Hf, U, Th, Eu, Sr, Ba and Cs). Rare Earth Elements (REEs) display a different behavior: (i) they show the lowest RVC, down to 0.04 and (ii) their RVC decreases systematically with ionic radius passing from light to heavy REEs. Among REEs, Eu shows peculiar behavior in that its RVC is the highest and is of the same order of magnitude of elements such as Th or Zr.

Similar to numerical simulations (Chap. 4), the Shannon entropy (S_N) was estimated using the method presented in the previous chapter. The graph in Fig. 5.8 reports the variation of the normalized Shannon entropy (S_N) against mixing time (t) for major elements. As for the case of numerical simulations, S_N increases with increasing mixing time for all elements. At longer mixing times (i.e. 120 min) S_N relaxes towards constant values. The different elements display varied slopes (dS/dt) of the linear part of the curves.

Figure 5.9 shows the variation of Shannon entropy (dS/dt) against RCV. As for the case of numerical simulations, the plot shows a linear positive correlation between these two variables. In particular, the highest values of dS/dt are recorded

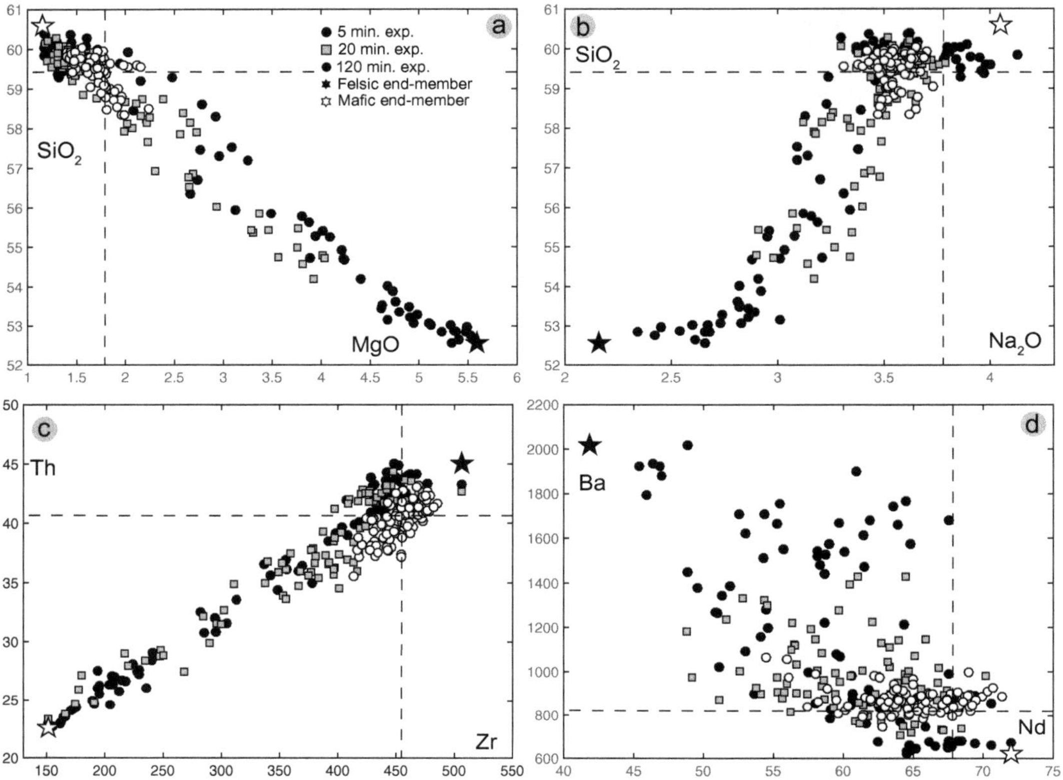

Fig. 5.5 Binary inter-elemental plots of some representative elements analysed along the transects displayed in Fig. 5.2. The two dashed lines intersect the x and y axes at the concentration of the hybrid concentration and is the focal point where the homogeneous system should collapse at infinite mixing time

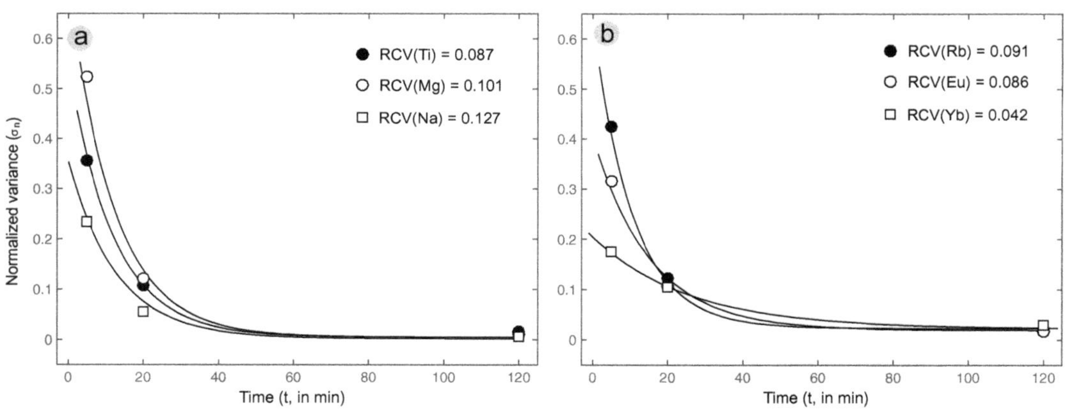

Fig. 5.6 Relaxation of Concentration Variance (RCV) for some representative **a** major and **b** trace elements. RCV (i.e. parameter R of the fitting in Eq. 4.6) is reported for each element

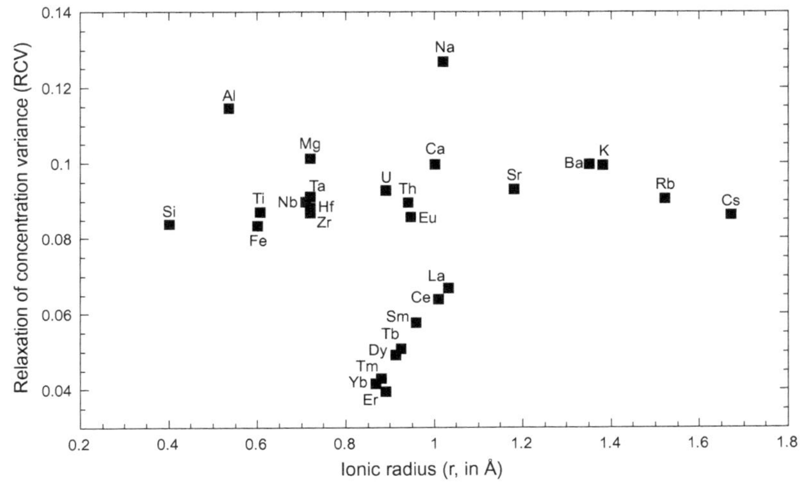

Fig. 5.7 Estimated values of Relaxation of Concentration Variance (RCV) for chemical elements as a function of ionic radius. Ionic radius data from (Shannon 1976)

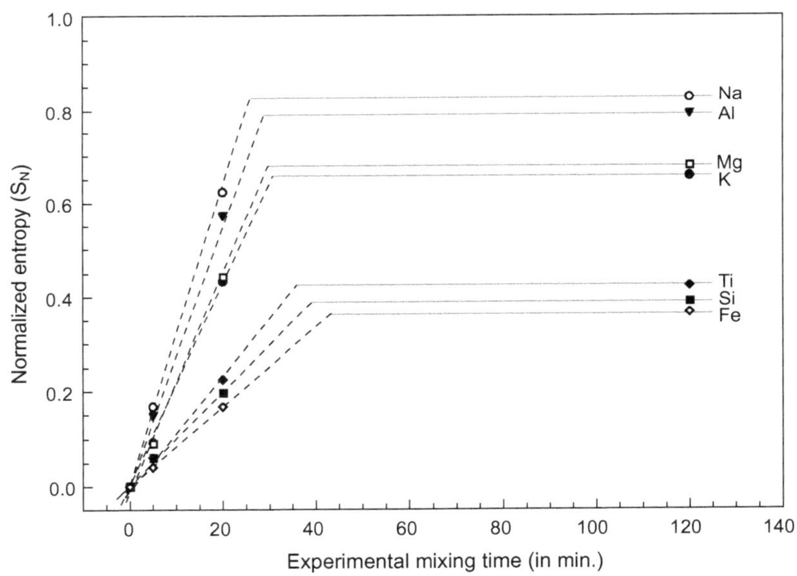

Fig. 5.8 Variation of Shannon entropy plotted against mixing time (t). The linear portion of the data set was fitted by linear regression (dashes lines)

for Na followed by Al, Mg, K, Ca, Ti, Si and Fe. This sequence matches perfectly the sequence of element mobility defined on the basis of the RCV method (e.g. Perugini et al. 2013). The rate of increase of Shannon entropy is different for the different chemical elements, each having different mobility in the mixing system. In detail, results indicate that elements having larger mobility display also a larger rate of production of hybrid compositions.

Results from these experiments highlighted a good matching between numerical and physical models. For example, there is a close matching between the evolution of compositional histograms, concentration variance, and Shannon entropy from numerical simulations and experiments. Experiments also allowed us to move further steps in the understanding of element mobility, as shown by the relationship derived between RCV and the ionic radius of elements. In this context it is clear that numerical simulations and experiments are complementary approaches that might allow us to draw the most complete picture about the complexity of magma mixing. However, although

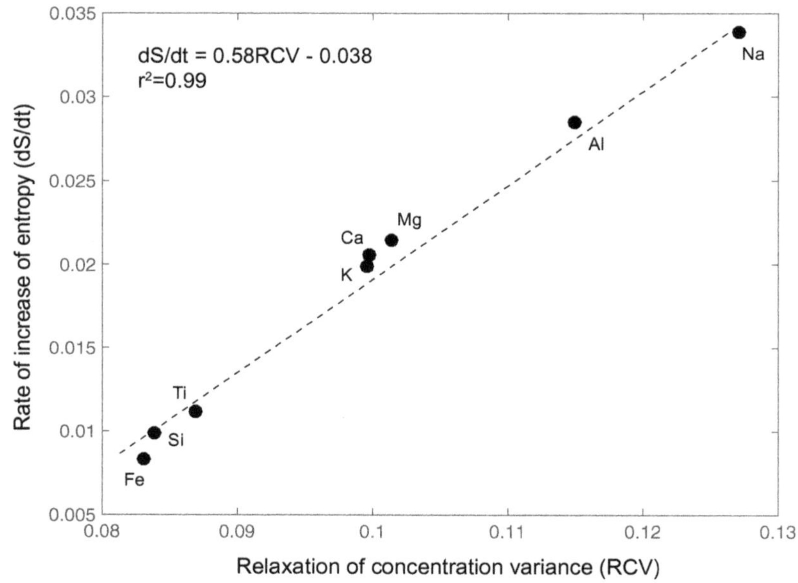

Fig. 5.9 Variation of the Relaxation of Concentration Variance (RCV) plotted against dS/dt. Best-fit line and linear equation are also displayed in the graph

the experiments presented above represent an excellent launch pad for a more detailed comprehension of this process, we still lack precise control of the very basic processes triggering the mutual dispersion of magmas, i.e. stretching and folding. As this process is the pre-requisite for magma mixing it must be better understood. For this reason, further experimental simulations can be designed. This is discussed in the next section.

5.3 The Chaotic Magma Mixing Apparatus

The need to tackle more in detail the problem of experimental magma mixing required the construction of a new device. Here we report on a new apparatus that was designed to perform experiments with high-viscosity (up to 10^6 Pa s) and high-temperature (up to 1300 °C) natural silicate melts. The device will allow us to follow in time and space the evolution of stretching and folding processes and the associated modulation of chemical composition. A prototype of this experimental setup was first developed at the Department of Earth and Environmental Sciences, Ludwig-Maximilians-University (LMU) of Munich (Germany), where a number of experiments were performed (e.g. Cristina and de Campos et al. 2011; Perugini et al. 2012; Morgavi et al. 2013). This prototype represented a fundamental step towards the aim of reproducing magma mixing processes. A few years later a new experimental setup was developed at the Department of Physics and Geology of the University of Perugia (Italy). This was termed Chaotic Magma Mixing Apparatus (hereafter COMMA, Morgavi et al. 2015). The new experimental system can accommodate voluminous samples (up to 75 cm^3 of melt). An advantage of this setup is that it permits the insertion of thermocouples close to the experimental sample to measure the melt temperature during the course of the experiments. The COMMA is inspired by the Journal Bearing Flow (JBF) system, a device used to study mixing of low viscosity (compared to natural silicate melts) and low temperature liquids (e.g. Swanson and Ottino 1990; Galaktionov et al. 2002).

A JBS consists of an outer cylinder, filled with the fluids of interest, and an off-centre inner cylinder (Fig. 5.10). The geometry of the system is determined by two parameters: (a) the ratio of the radii of the two cylinders, $r = R_{in}/R_{out} = 1/3$, and (b) the eccentricity ratio to the outer cylinder $\epsilon = \delta/R_{out}$, where δ is the distance between the centres of the inner and outer cylinders (R_{in} and

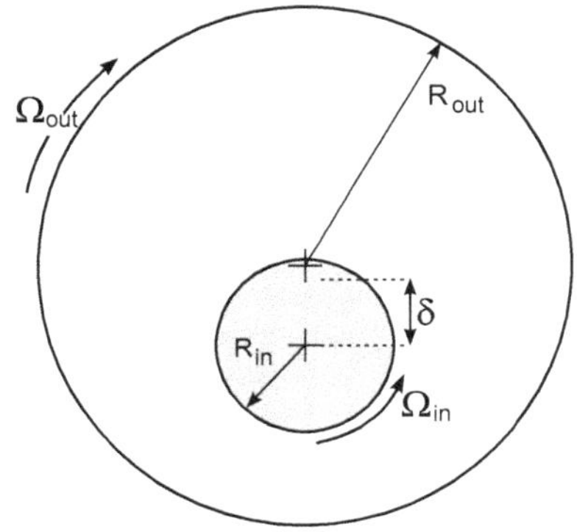

Fig. 5.10 Top-view of the geometry of the chaotic magma mixing apparatus (COMMA)

R_{out}). Chaotic mixing is triggered by the rotation of the cylinders in opposite directions (e.g. Swanson and Ottino 1990; Galaktionov et al. 2002).

The COMMA apparatus was designed to achieve three main targets: (1) allowing independent rotations of the two cylinders, (2) positioning the device in a high-temperature furnace, and (3) cooling very rapidly the melts after the experiments, in order prevent crystallization of the samples.

The COMMA is shown in Fig. 5.11. Thermal ramps are precisely controlled via Personal Computer. On the top and the bottom ends of the furnace are positioned cooling heads in order to prevent overheating of the motors.

Two pneumatic cylinders are used for the vertical movement of the furnace and this allows to prepare with precision the geometry of the experiments. This vertical movement also give the opportunity to expose rapidly to ambient temperature the experimental change (cooling rates of ca. 50 °C/min), avoiding crystallization of the sample. In order to rotate independently the outer and inner cylinders, two motors are positioned above and below the furnace (Fig. 5.11). The motors can be precisely moved in the horizontal and vertical position to achieve the geometry required in the experiments.

An Al_2O_3 rod, acting as the inner cylinder, is fixed to the upper motor. A second alumina rod is fixed to the lower motor and holds the outer cylinder (Fig. 5.11). The lower end of the first rod is covered by a Pt_{80}-Rh_{20} foil in order to avoid contamination of the sample with alumina during the experiments. S-type thermocouple wires are passed through the upper Al_2O_3 rod and allow measuring the temperature in the experimental sample. A Pt_{80}-Rh_{20} crucible (i.e. the outer cylinder) is fixed on the top of end of the alumina rod fixed to the lower motor. This configuration allows us to perform chaotic magma mixing experiments with good precision.

At the end of the experiments, glasses are cored out from the crucible and set in epoxy. This way they can be sliced and analysed.

The choice of reproducing the journal bearing flow geometry to perform magma mixing experiments is motivated by the fact that experiments can be carefully planned using numerical simulations. The analytical solution for this system allows to model it for any combination of rotation and position of the two cylinders (e.g. Ottino and Ottino 1989; Wannier 1950). This is a crucial step because we can evaluate the behaviour of the mixing system before performing a physical experiment.

Chaotic magma mixing experiments were performed using the system configuration reported in Muzzio et al. (1992). This consists of alternating rotations of the outer and inner cylinders with the following scheme: (1) two complete rotations of the outer cylinder; (2) six complete rotations of the inner cylinder. This protocol was repeated twice.

Figure 5.12 reports the outcome of the numerical simulation at different times. Here we can follow in time the progression of the mixing process, evaluate the amount of deformation of the two melts, and observe how flow fields generate a complex pattern of filaments.

A first pilot experiment was performed at the Department of Earth and Environmental Sciences, Ludwig-Maximilians-University of Munich (Germany) using as end-members a peralkaline haplogranite melt (hereafter Melt A) (Dingwell et al. 1998), and a haplobasaltic melt composition (hereafter Melt B), corresponding to the 1-atm eutectic composition of the anorthite–diopside (An–Di) binary system. Further experiments performed using natural silicate melts will be discussed in the next chapters when dealing with the interpretation of natural magma mixing patterns. The above

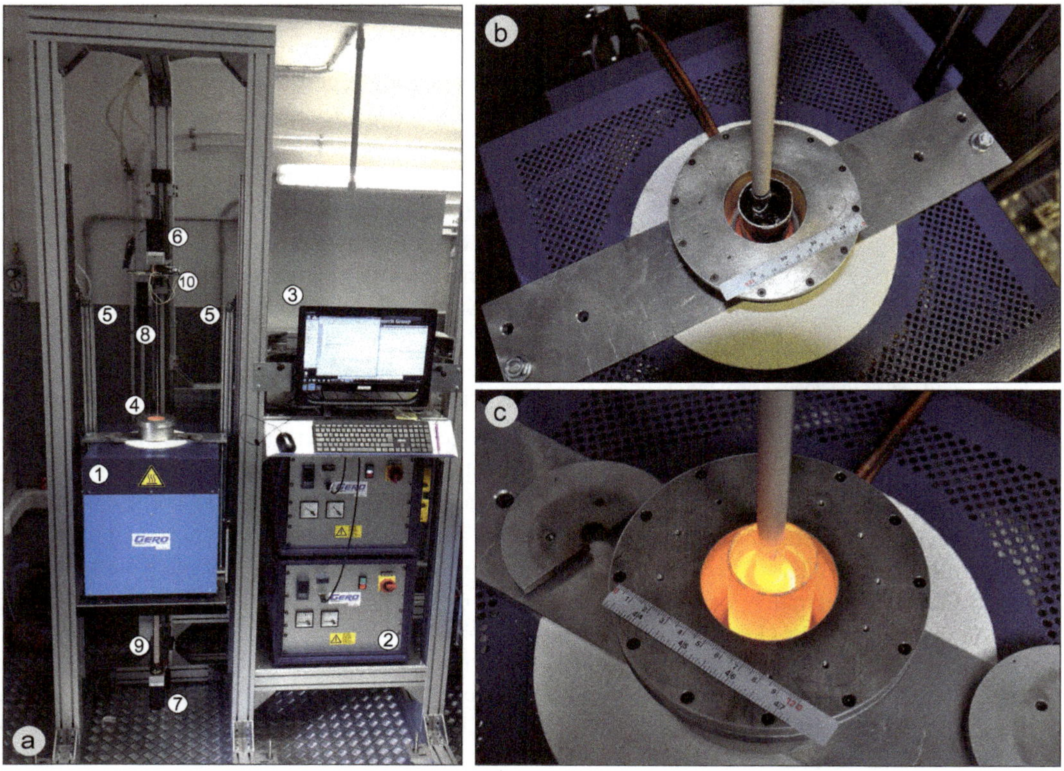

Fig. 5.11 **a** The chaotic magma mixing apparatus (COMMA): (1) furnace; (2) furnace control unit; (3) Personal Computer controlling temperature of the furnace, direction of rotation and velocity of the motors; (4) cooling head; (5) pneumatic pistons for the vertical positioning of the furnace; (6) upper motor; (7) lower motor; (8) Al_2O_3 rod acting as inner cylinder (covered by a Pt foil in the lower part); (9) square-section Al_2O_3 rod sustaining the outer cylinder inside the furnace; and (10) wireless thermocouple to measure the temperature in the experimental sample. **b** and **c** Zoomed-in views of the upper part of the COMMA showing the configuration of the mixing system before and after completion of the experiment, respectively

mentioned two compositions can be considered, as a first approximation, as proxies for mafic and felsic natural melts. The viscosity of the two starting melts was calculated at experimental conditions (i.e. 1400 °C) using the model reported in Giordano et al. (2008). Melt A and B result to have viscosity of ca. 1550 Pas and 1.4 Pas, respectively. This corresponds to a viscosity ratio of 1.1×10^3. The density of the melts was estimated using the model of Lange and Carmichael (1987) and resulted to be 2.26 and 2.52 g/cm^3 for Melt A and B, respectively. Volume proportions of the felsic and mafic melt were 65.0% and 35%, respectively. The experimental protocol was the same as for the numerical simulation shown in Fig. 5.12. The reader can find further details in Perugini et al. (2012).

The Reynolds number during the experiment was:

$$\mathrm{Re} = \frac{\rho V L}{\mu} \tag{5.3}$$

where V is characteristic velocity, $V = |V_{in}| - |V_{out}$, with $V_{in} = \Omega_{in} R_{in}$ and $V_{out} = \Omega_{out} R_{out}$; L is a characteristic length $L = R_{out} - R_{in}$; ρ and μ are the density and viscosity and Ω_{in} and Ω_{out} are the angular velocities of inner and outer cylinder, respectively (e.g. Ottino and Ottino 1989). Re during the experiment was of the order of 10^{-7}.

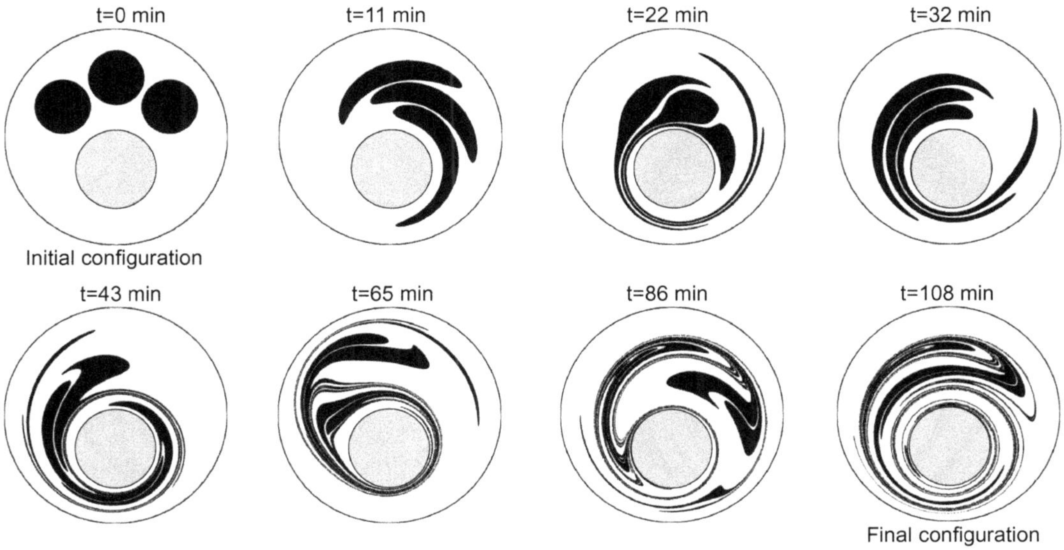

Fig. 5.12 Numerical simulation of the mixing system. The initial configuration (top-left corner) evolves in time through the repetition of stretching and folding dynamics leading to the final configuration after 108 min. Modified from Perugini et al. (2012)

Figure 5.13b displays a section of the experiment sliced orthogonal to the long axis. Figure 5.13a shows the mixing pattern resulting from the numerical simulation. Comparison of the two pictures indicates an almost perfect matching between the patterns generated by the experiment and those resulting from the numerical simulation.

The compositional variability across the mixing patterns was measured along some transects by EMPA and on areas by X-ray elemental mapping.

The variation of chemical elements across two transects is shown in Fig. 5.14. In general, element concentration oscillates depending on the relative occurrence of filaments of the two melts. The transition between filaments is continuous indicating that chemical diffusion occurred among the filaments. The variations of SiO_2 and MgO display specular patterns. Na_2O shows a different pattern, suggesting a different mobility of this element in the mixing system.

The evolution of the mixing process is also shown in the binary plots of Fig. 5.15. The plots show complex patterns that deviate significantly from a simple linear mixing relation.

The two-dimensional spatial variability of concentration fields is shown in Fig. 5.16. For each area the variation of CaO and Na_2O is displayed. The compositional variation for these elements is reported as $3D$-graphs, where x and y represent the coordinates of the mapped area and z represents concentration. The compositional variations produced by mixing is visible as alternations of ridges and valleys depending on the relative abundance of the melts. As in the case of transects, the compositional variation of one element does not overlap with the variation of another (e.g. CaO and N_2O in the case of Fig. 5.16).

It is therefore evident that the different chemical elements display different mobility during the mixing process, as already observed for the centrifuge experiments. This issue is of

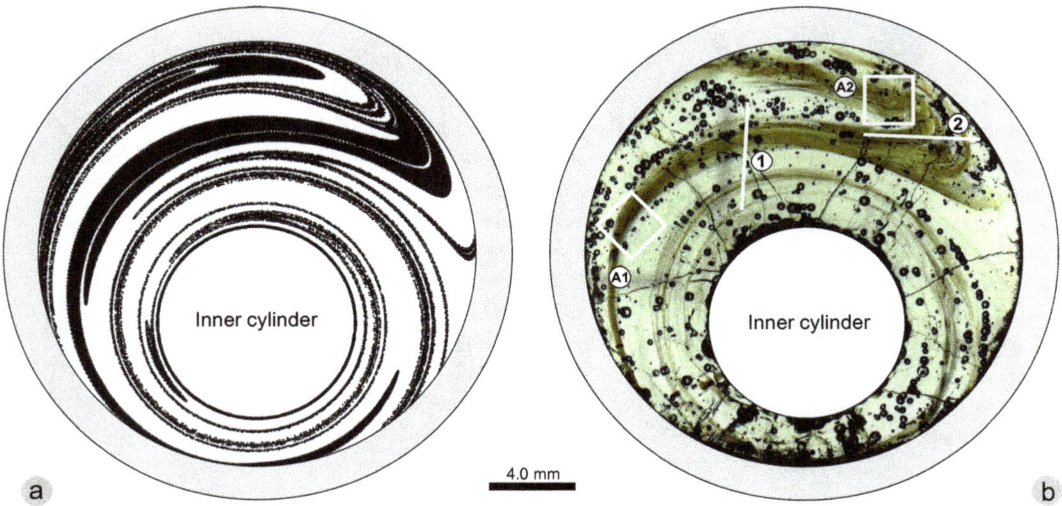

Fig. 5.13 Comparison between the mixing patterns reproduced numerically (**a**) and the outcome of the magma mixing experiment (**b**). Dark spots in **b** are vesicles formed by air entrapment during the preparation of the experimental sample. The transects (1 and 2) and the areas (A1 and A2) on which geochemical analyses were performed are also shown in **b**

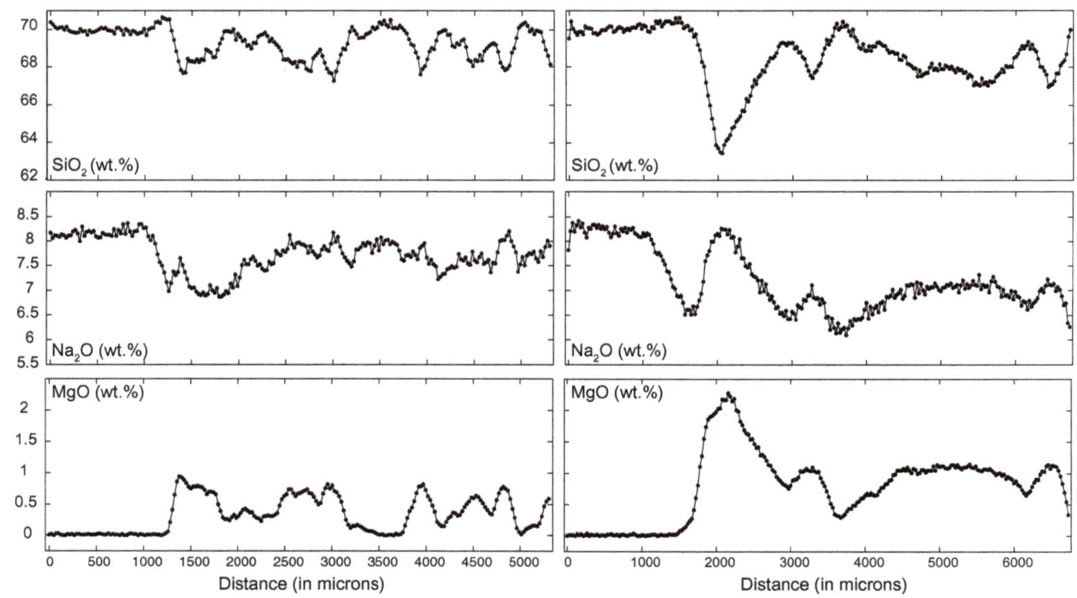

Fig. 5.14 Variation of concentration of representative chemical elements along the analysed transects. Left column: transect n. 1; right column: transect n. 2 (see Fig. 5.13b)

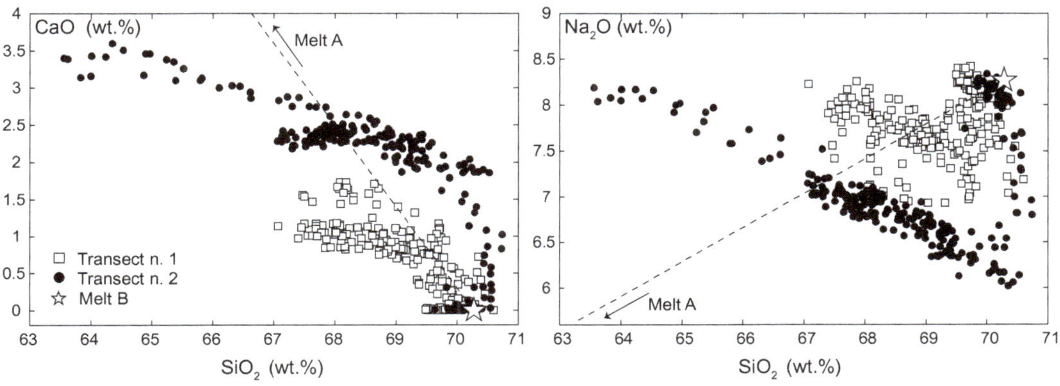

Fig. 5.15 Representative binary plots displaying the compositional variability produced by the mixing process. The mixing line connecting the two end-members is also shown

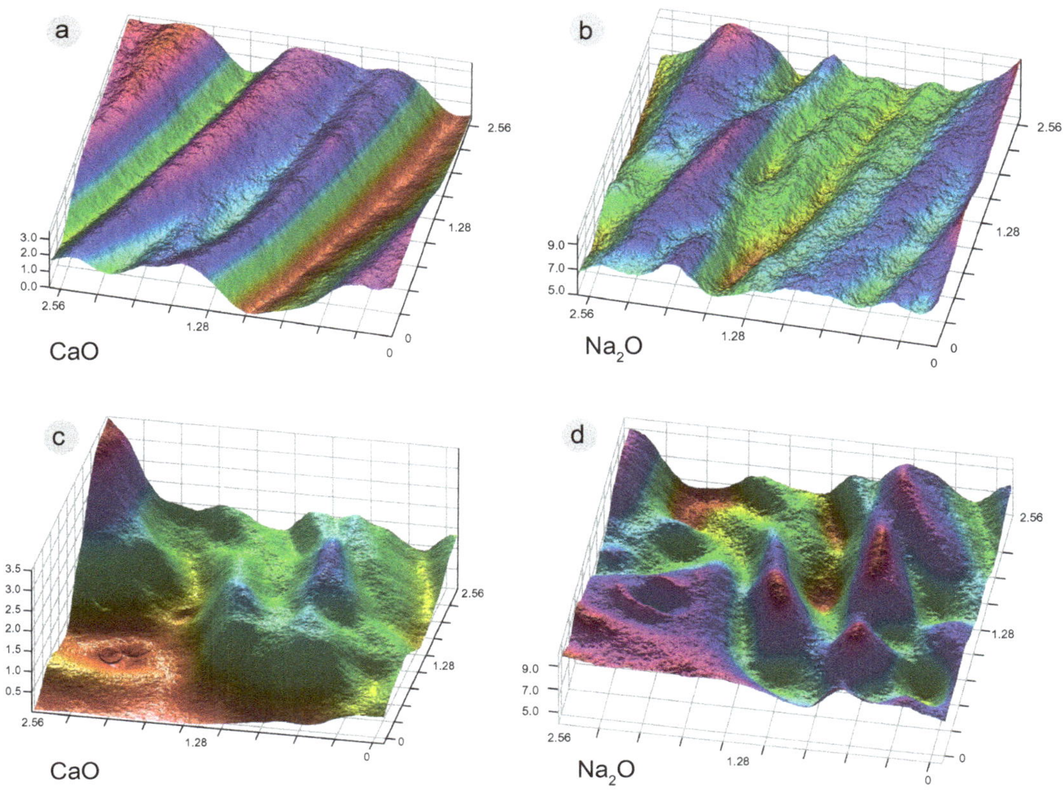

Fig. 5.16 Compositional variability of CaO and Na_2O reported as $3D$-graphs. The z coordinate reports element concentration. Imaged areas of the mixing system are indicated by the white squares in Fig. 5.13b

paramount importance for both the petrological and volcanological implications of magma mixing and will be discussed to a greater extent in the next chapters.

In this and the previous chapter we have evaluated how numerical simulations and experiments can aid in understanding the complex evolution in time and space of magma mixing. It is now time to move to Nature and apply what we have learned until now to interpret the structural and compositional mixing patterns in igneous rocks.

References

Arienzo I et al (2009) Isotopic evidence for open system processes within the Campanian Ignimbrite (Campi Flegrei-Italy) magma chamber. Bull Volcanol 71(3):285–300

Cristina CPP, de Campos P et al (2011) Enhancement of magma mixing efficiency by chaotic dynamics: an experimental study. Contribut Miner Petrol 161(6):863–881. https://doi.org/10.1007/s00410-010-0569-0

De Campos CP et al (2004) Decoupled convection cells from mixing experiments with alkaline melts from Phlegrean Fields. Chem Geol 213(1–3):227–251. ISSN: 00092541. https://doi.org/10.1016/j.chemgeo.2004.08.045

de Vita S et al (1999) The Agnano-Monte Spina eruption (4100 years BP) in the restless Campi Flegrei caldera (Italy). J Volcanol Geotherm Res 91(2–4):269–301

Di Vito MA et al (1999) Volcanism and deformation since 12,000 years at the Campi Flegrei caldera (Italy). J Volcanol Geothermal Res 91(2–4):221–246

Dingwell DB, Hess KUKU, Romano C (1998) Extremely fluid behavior of hydrous peralkaline rhyolites. Earth Planet Sci Lett 158(1–2):31–38

Dorfman A, Hess KU, Dingwell D (1996) Centrifuge-assisted falling-sphere viscometry. Eur J Min 8(3):507–514

Galaktionov OS, Anderson PD, Peters GWM (2002) Structure development during chaotic mixing in the journal bearing flow. Phys Fluids 14(9):3009–3017

Giordano D, Russell JK, Dingwell DB (2008) Viscosity of magmatic liquids: A model. Earth Planet Sci Lett 271(1–4):123–134

Kouchi A, Sunagawa I (1985) A model for mixing basaltic and dacitic magmas as deduced from experimental-data. Contribut Min Petrol 89(1):17–23

Kouchi A, Sunagawa I (1983) Mixing basaltic and dacitic magmas by forced convection. Nature 304(5926):527–528. ISSN: 00280836. https://www.nature.com/articles/304527a0. https://doi.org/10.1038/304527a0

Lange RA, Carmichael ISE (1987) Densities of Na2o-K2o-Cao-Mgo-Feo-Fe2o3-Al2o3-Tio2-Sio2 liquids - new measurements and derived partial molar properties. Geochimica et Cosmochimica Acta 51(11):2931–2946

Laumonier M et al (2014) On the conditions of magma mixing and its bearing on andesite production in the crust. Nat Commun 5(1):1–12. ISSN: 20411723. https://doi.org/10.1038/ncomms6607

Laumonier M et al (2015) Experimental mixing of hydrous magmas. Chem Geol 418:158–170. ISSN: 00092541. https://doi.org/10.1016/j.chemgeo.2015.10.031

Morgavi D et al (2013) Interactions between rhyolitic and basaltic melts unraveled by chaotic mixing experiments. Chem Geol 346:199–212. ISSN: 00092541. https://doi.org/10.1016/j.chemgeo.2012.10.003

Morgavi D et al (2015). High-temperature apparatus for chaotic mixing of natural silicate melts. https://doi.org/10.1063/1.4932610

Muzzio FJ, Swanson PD, Ottino LM (1992) Mixing distributions produced by multiplicative stretching in chaotic flows. Int J Bifurcat Chaos 2:37–50

Ottino JM, Ottino JM (1989) The kinematics of mixing: stretching, chaos, and transport

Perugini D et al (2004) Kinematic significance of morphological structures generated by mixing of magmas: a case study from Salina Island (southern Italy). Earth Planet Sci Lett 222(3–4): 1051–1066. ISSN: 0012821X. https://doi.org/10.1016/j.epsl.2004.03.038

Perugini D et al (2008) Trace element mobility during magma mixing: preliminary experimental results. Chem Geol 256(3–4):146–157. ISSN: 00092541. https://doi.org/10.1016/j.chemgeo.2008.06.032

Perugini D et al (2012) The space and time complexity of chaotic mixing of silicate melts: Implications for igneous petrology. Lithos 155:326–340. ISSN: 00244937 (ISSN). https://doi.org/10.1016/j.lithos.2012.09.010

Perugini D et al (2013) Relaxation of concentration variance: A newtool to measure chemical element mobility during mixing of magmas. Chem Geol 335:8–23. https://doi.org/10.1016/j.chemgeo.2012.10.050

Swanson PD, Ottino JM (1990) A comparative computational and experimental-study of chaotic mixing of viscous fluids. J Fluid Mech 213:227–249

Wannier GH (1950) A contribution to the hydrodynamics of lubrication. Q Appl Math 8(1):1–32

Yuan LA, Zheng YF (2011) A microdevice for the mixing of a highly viscous biosample with water/membrane protein solution using microchannel and centrifugation. Jala 16(1):68–81

Zimanowski B, Büttner R, Koopmann A (2004) Experiments on magma mixing. Geophys Res Lett 31(9). ISSN:00948276. https://doi.org/10.1029/2004GL019687

Part III

Magma Mixing: A Petrological Process

The Beginning: Mafic Magmas Invading Felsic Magma Chambers

6

All scientific advance in discovering the laws of nature has been made by first simplifying a problem and reducing it to certain ideal conditions, and this is what mathematicians have done in studying the motion of a liquid.

Henry S. Hele-Shaw

Abstract

The initial stages of magma interaction, i.e. the moment at which the two magmas come into contact, is a crucial point since the fate of magma mixing and its development in space and time largely depend on the outcome of this first encounter. In this chapter we discuss the possible fluid dynamic and thermodynamic processes developing during the injection of mafic magmas into felsic magma chambers. We will combine the study of the complexity of contact morphologies between magmas observed in natural outcrops with fluid-mechanics experiments and discuss the results in terms of evolution of rheological contrasts during the injection process.

6.1 Introduction

Before discussing the development of widespread magma mixing processes in magma chambers, conduits and lava flows, it is worth exploring the initial stages of magma interaction, i.e. the moment at which the two magmas come into contact. This is a crucial point since the fate of magma mixing and its development in space and time largely depend on the outcome of this first encounter.

Several attempts to study this initial configuration have been made mainly using fluid-dynamics and numerical models (e.g. Cardoso and Woods 1996; Campbell and Turner 1986; Sparks and Marshall 1986; Cruden et al. 1995). Nevertheless, the initial stages of intrusion of mafic magmas into felsic reservoirs are still poorly understood (e.g. Wiebe 1993; Waight et al. 2001). A possible cause for the scarcity of works on these crucial stages is that they are rarely preserved in nature.

In this chapter we discuss briefly the possible fluid dynamic and thermodynamic processes developing during the injection of mafic magmas into felsic magma chambers. We will combine the study of the complexity of contact morphologies between magmas observed in natural outcrops with fluid-mechanics experiments and try

D. Perugini, *The Mixing of Magmas*, Advances in Volcanology,
https://doi.org/10.1007/978-3-030-81811-1_6

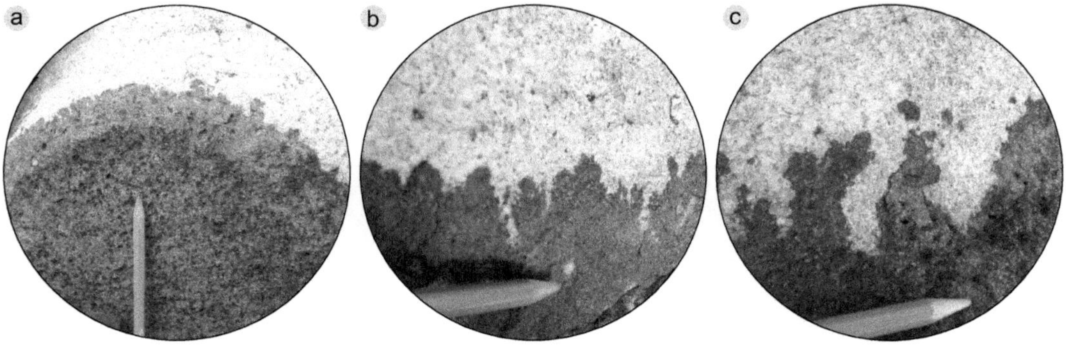

Fig. 6.1 Representative mafic-felsic contact interfaces in the Vegetation Unit outcrop

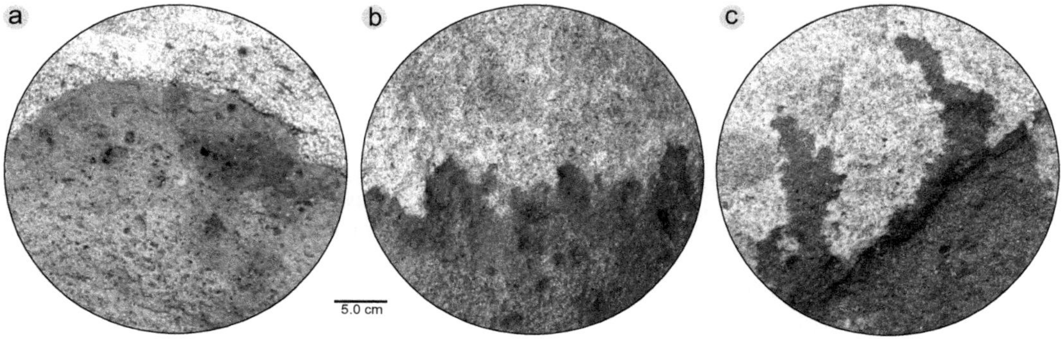

Fig. 6.2 Representative mafic-felsic contact interfaces in the Pleasant Bay outcrop

to discuss the results in terms of evolution of rheological contrasts during the injection process.

6.2 Natural Outcrops

Superb examples of the initial stages of the invasion of mafic magmas in a felsic magmas are well preserved in granitoid rocks. Three natural case studies will be considered here.

The first outcrops is represented by the intrusive bodies of Vegetation Unit (Terra Nova Intrusive Complex, Antarctica; e.g. Di Vincenzo and Rocchi 1999; Perugini and Poli 2005). The Vegetation Unit shows the interaction between a monzodiorite and a granite. This produced pillowing of the mafic magma in the felsic one generating strongly variable contact morphologies (Fig. 6.1).

The second case study is represented by the Pleasant Bay intrusive body (Maine, USA; e.g. Wiebe 1993; Hogan and Sinha 1989). Here, the effects of magma interaction between gabbro and granite magmas have been documented (Wiebe 1993; Waight et al. 2001). As for the case of Vegetation Unit, the contact morphology between the two magmas are very varied (Fig. 6.2).

The third case study is the Aztec Wash pluton (Eldorado Mountains (Nevada; e.g. Howard and John 1987; Faulds et al. 1990). Here the granitoid mass was intruded by a mafic magma of monzodiorite composition (e.g. Falkner et al. 1995). As in the other cases, the injection of the mafic magma produced varying contact interfaces between the two magmas (Fig. 6.3).

In the three natural outcrops the morphology of the contacts between magmas ranges from ca. rounded to very irregular. In the latter case, fingers of the mafic magma penetrate and displace the felsic host. The fact that we can observed qualitatively the same variability of mafic-felsic interfaces indicates that similar processes might have developed. Our task in this chapter is to attempt

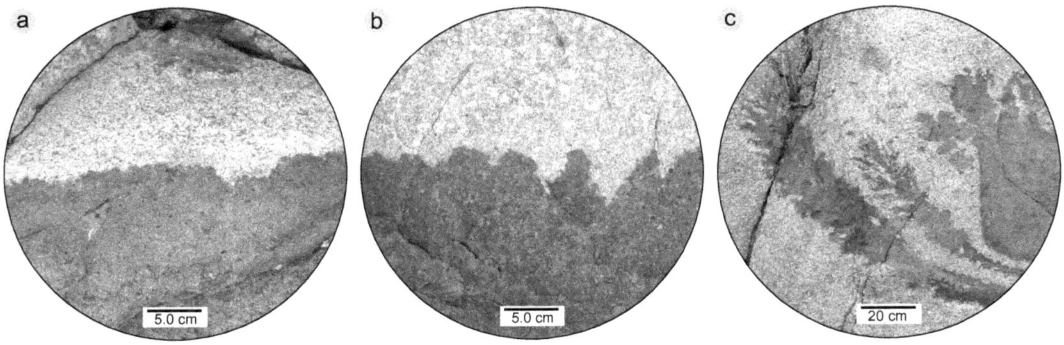

Fig. 6.3 Representative mafic-felsic contact interfaces in the Aztec Wash outcrop

to decipher the information preserved in these patterns.

As the mafic/felsic interface appears to be a key feature, its analysis may help to decipher the processes acting during the initial stages of injection of mafic magmas into the felsic ones.

6.3 Fractal Analysis of Mafic-Felsic Interfaces

As we have seen in Chap. 2, fractal dimension can measure complex morphologies. We can, therefore, apply fractal analysis to quantify the variability of mafic/felsic interfaces starting from outcrops pictures.

As we have seen in Chap. 2 one of the most useful methods to measure the fractal dimension is the box-counting technique (see Chap. 2 for details). Before applying this method, however, pictures need to be converted to binary (black and white) images. This can be done by thresholding grayscale images to produce pictures in which the interface between mafic and felsic magmas is replaced by a black line. Once the contact interface is extracted from outcrop pictures, the box-counting analysis can be performed. This process is shown in Fig. 6.4.

Figure 6.4c shows an example of the application of the box-counting technique to a representative mafic-felsic interface from the Vegetation Unit outcrop (Fig. 6.4a, b). Data points follow a linear distribution, indicating that interfaces are fractals.

The technique was applied to all structures from the three outcrops. Figure 6.5 shows frequency histograms of D_{box} values for the three case studies. Although the number of samples is different for the different outcrops, the shape of the histograms is very similar: in the three outcrops D_{box} values range from ca. 1.0 to 1.7, indicating a considerable variability of interface morphologies.

6.4 Fluid-Mechanics Experiments

The morphologies of mafic-felsic contacts observed in natural outcrops match those formed by injecting a low-viscosity fluid into a high-viscosity fluid (e.g. Nittmann et al. 1985). This process is known as "viscous fingering" and was first studied by Saffman and Taylor (1958) using the Hele-Shaw cell. An Hele-Shaw cell consists of two parallel plates separated by a narrow gap. When the low-viscosity fluid displaces the high-viscosity one, the so-called Saffman-Taylor instability occurs, and the contact interface evolves into a finger-like pattern. The morphology of contacts between the two fluids depends on the viscosity ratio (V_R) ($= \eta_1/\eta_2$), where η_1 and η_2 are the viscosity of the high and low viscosity fluid, respectively. For values of V_R close to unity the interface is roughly circular. As V_R increases, finger-like fractal patterns are formed.

Viscous fingering experiments were performed by injecting water-glycerin solutions into pure glycerin at different viscosity ratios. The range of

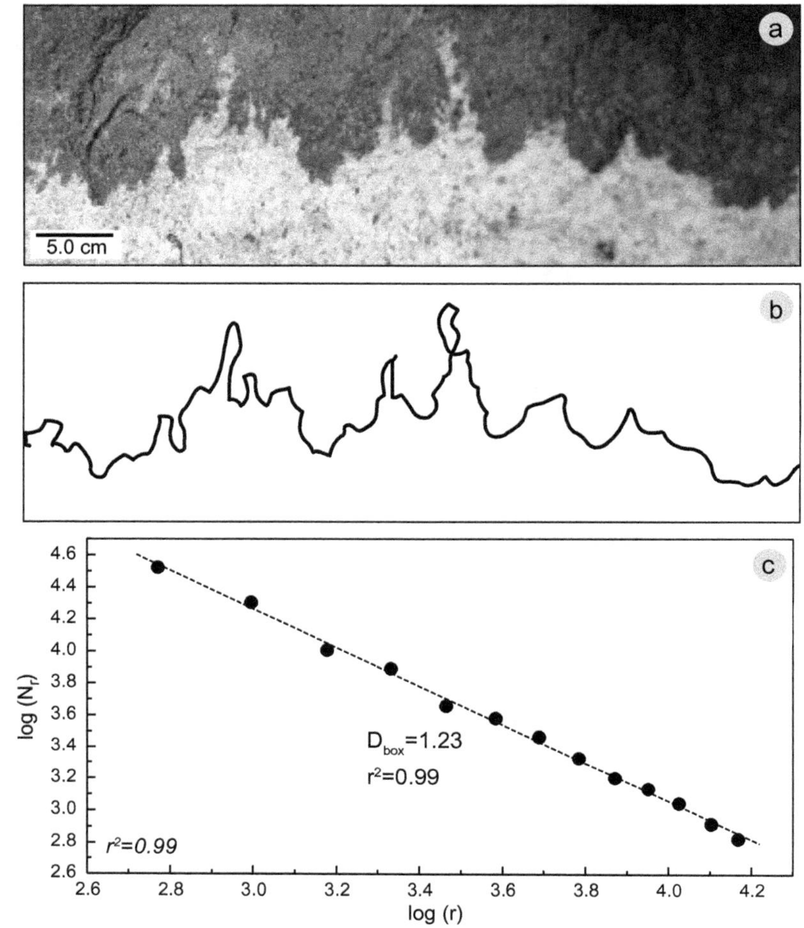

Fig. 6.4 Procedure used to measure fractal dimension (D_{box}) of mafic/felsic interfaces; **a** representative image of mafic-felsic morphology; **b** contact interface traced from **a** and used to estimate the fractal dimension (D_{box}) with the box-counting method; **c** estimate of D_{box} by linear interpolation of log(r) versus log(N_r) graph; slope of linear interpolation is equal to $-D_{box}$

viscosity ratios (V_R) that can be covered using water ($\eta_2 = 1.005$ mPas at 20 °C) and glycerin ($\eta_1 = 1410$ mPas at 20 °C) is from ca. 3.0 to 1400. The injected fluid was dyed to highlight fluid interfaces. Figure 6.6 shows a sketch of the Hele-Shaw cell used in the experiments. It consists of two Plexiglas plates separated by thin gap. The injection of the fluid occurs through a hole placed at the centre of the upper plate. The cell was first filled with pure glycerin. Then, the low-viscosity dyed solution (obtained by mixing different proportions of water and glycerin) was injected.

Figure 6.7 displays the outcomes of some representative experiments. The figure shows that for viscosity ratios close to unity, the interface between the two fluids is circular (Fig. 6.7a); slightly increasing V_R, the low-viscosity fluid tends to penetrate the higher-viscosity one forming small bumps (Fig. 6.7b). This is the beginning of Saffman-Taylor instability. As the viscosity ratio progressively increases (Fig. 6.7c–d) finger-like patterns splitting into several lobes are formed.

Natural and experimental structures are very similar in that there is a large variability of morphologies ranging from rounded to fingered.

D_{box} for experimental viscous fingering structures was measured, as for natural structures. Results show that fractal dimension varies from 1.01 to 1.63, covering most of D_{box} values obtained for the natural patterns.

Figure 6.8 displays the variation of D_{box} as a function of the logarithm of viscosity ratio [log(V_R)]. In the graph are also reported values of fractal dimension D_{box} of viscous fingering

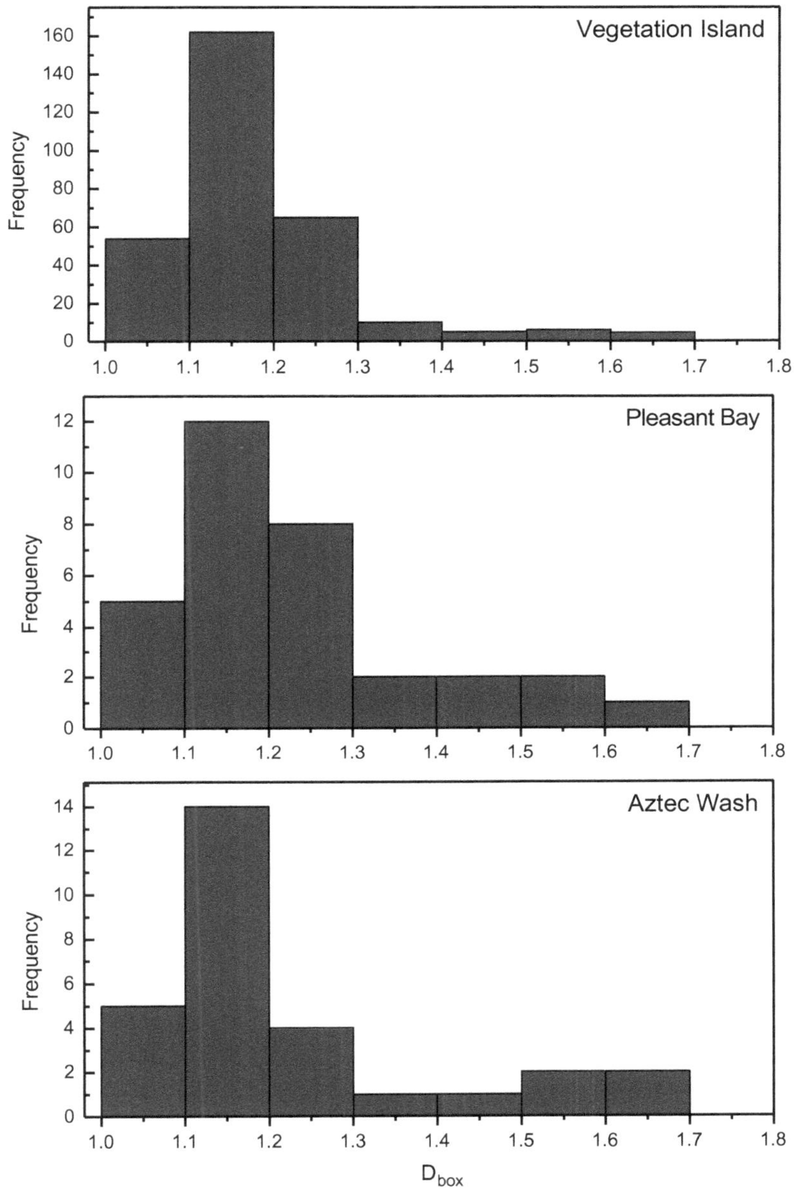

Fig. 6.5 Frequency histograms of fractal dimension (D_{box}) of contact interfaces in the three natural case studies

patterns from Allen and Boger (1988). Using these values allows us to extend the range of V_R that are regarded plausible during the injection of mafic magmas in felsic magmas (i.e., $10 \leq V_R \leq 10^4$; e.g. Snyder and Tait 1995). Figure 6.8 shows that data point can be fitted by an exponential relationship ($r^2 = 0.99$) of the form:

$$\log(V_R) = 0.013e^{3.34D_{box}} \tag{6.1}$$

This allows us to derive an experimental relationship between fractal dimension (D_{box}; i.e. a morphological descriptor) and viscosity ratio, V_R (i.e. a rheological descriptor). An important point is the appropriateness of the comparison between fractal dimension of natural and experimental structures. While natural structures are sections of $3D$ morphologies, experimental structures are essentially the result of $2D$ experiments.

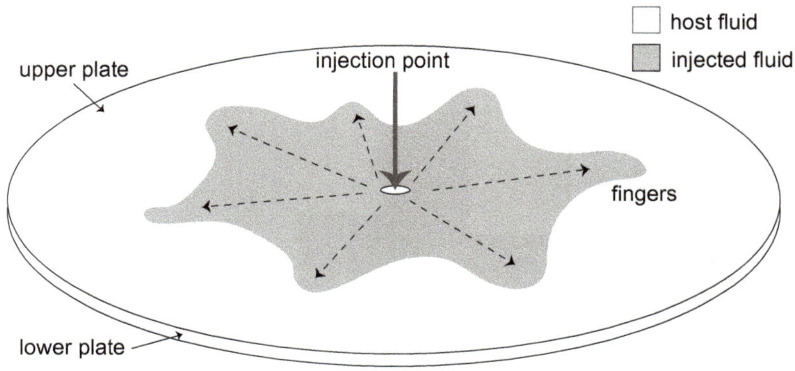

Fig. 6.6 Sketch of the Hele-Shaw cell used to perform viscous fingering experiments (additional details in Perugini and Poli 2005)

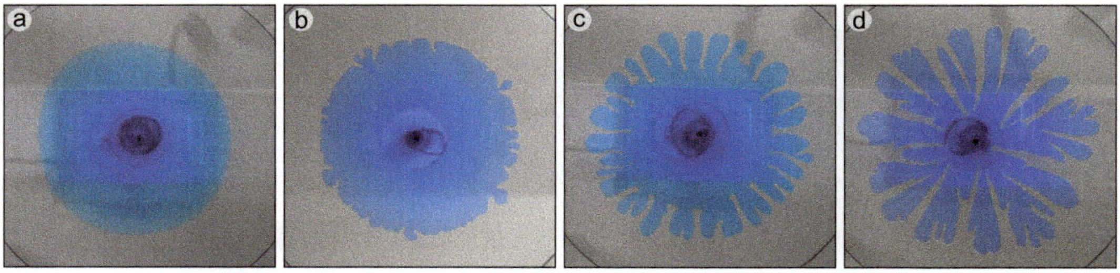

Fig. 6.7 Morphology of interfaces produced by viscous fingering experiments at progressively larger viscosity ratios (V_R); **a** $V_R = 1.0$, **b** $V_R = 2.69$, **c** $V_R = 12.9$, **d** $V_R = 1403$. Starting from **b** the onset of Saffman-Taylor instability can observed. This becomes more evident as the viscosity ratio increases **c**–**d**

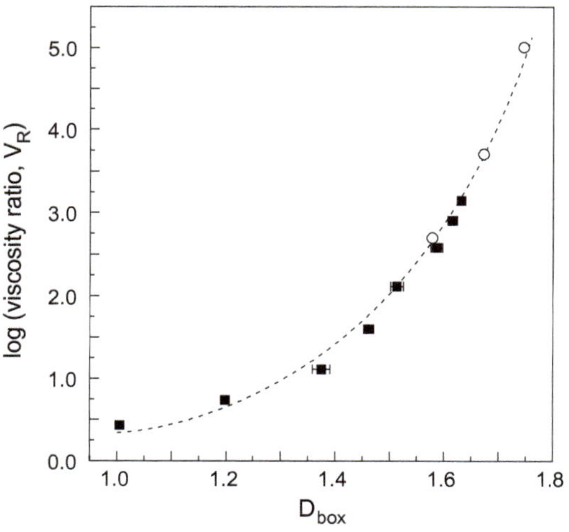

Fig. 6.8 Variation of D_{box} versus $\log(V_R)$ for viscous fingering structures generated experimentally. Open circles are experimental structures from Allen and Boger (1988); curve shows exponential fit (Eq. 6.1)

We can address this point, taking into account theoretical arguments (Mandelbrot 1982). Topologically, if S_1 and S_2 are two independent sets in a space with dimension d, and if codimension(S_1) $+$codimension$(S_2) < d$, the codimension of $S_1 \cap S_2$ is the sum of the codimensions of S_1 and S_2. Given a fractal set F in $3D$, which is cut by a plane, this means that the dimension of the set resulting from the intersection is one less than the dimension of F. More pragmatically, if the fractal dimension of a $3D$ structure (D_s) is 2.28, the fractal dimension of any intersecting plane is $D_I = 2.28 - 1 = 1.28$. Thus, different sections of the same $3D$ structure should have the same fractal dimension. These arguments have been experimentally proven in the case of turbulent jets (e.g. Sreenivasan et al. 1989), indicating their applicability to the real world.

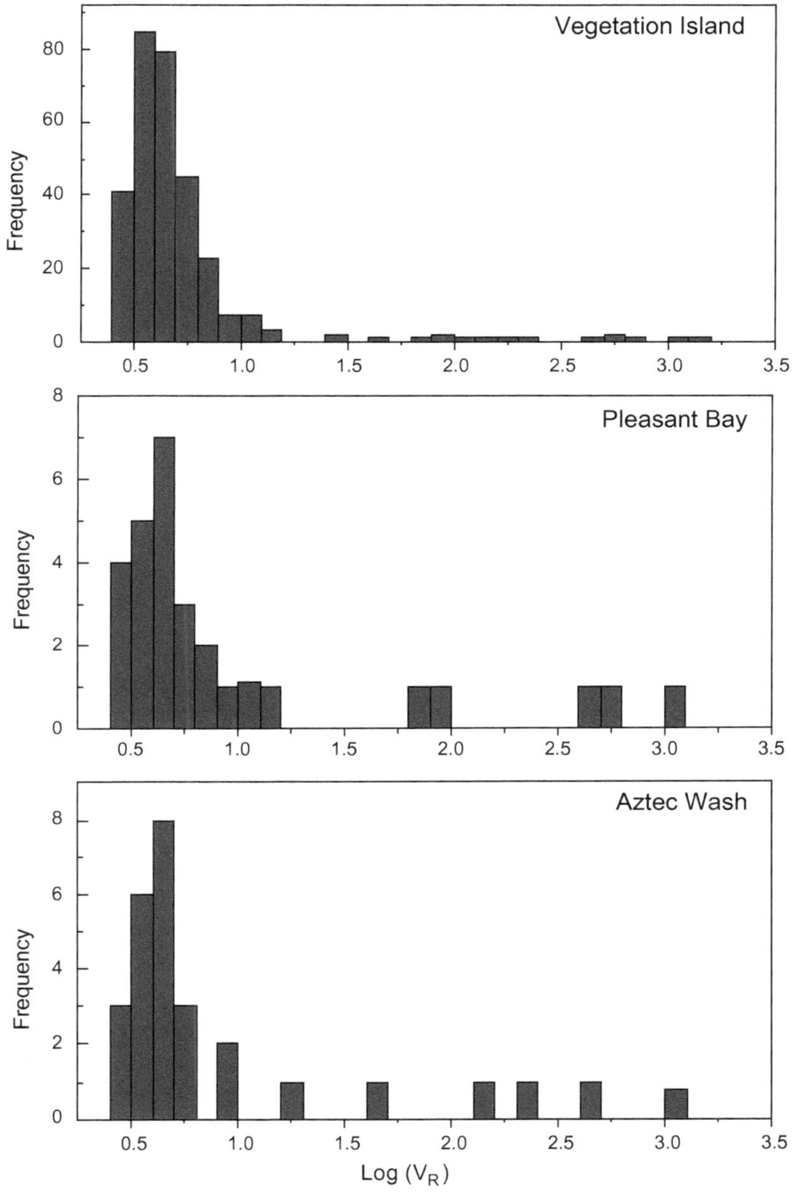

Fig. 6.9 Frequency histograms of $\log(V_R)$ values for the three natural case studies calculated using the empirical relationship derived from fluid-dynamics experiments (Eq. 6.1)

$\log(V_R)$ between magmas was calculated (Eq. 6.1) using the measured values of D_{box} for mafic-felsic interfaces. Figure 6.9 shows $\log(V_R)$ frequency histograms of D_{box} for the three natural outcrops and indicates that $\log(V_R)$ ranges from 0.4 (i.e., $V_R \approx 2.50$) to 3.12 (i.e., $V_R \approx 1328$). This implies that volumes of magmas with a large variability of V_R were present in the three studied magmatic systems. Most of $\log(V_R)$ values are in the range 0.4–0.6, corresponding to viscosity ratios (V_R) from ca. 2.50 to 4.00, for the three case studies.

Several authors indicated that during the interaction of magmas with different viscosity, the viscosity ratio is destined to decrease in time (e.g. Sparks and Marshall 1986; Poli et al. 1996). In detail, the continuous injection of the mafic magma into the felsic one will produce a progressive

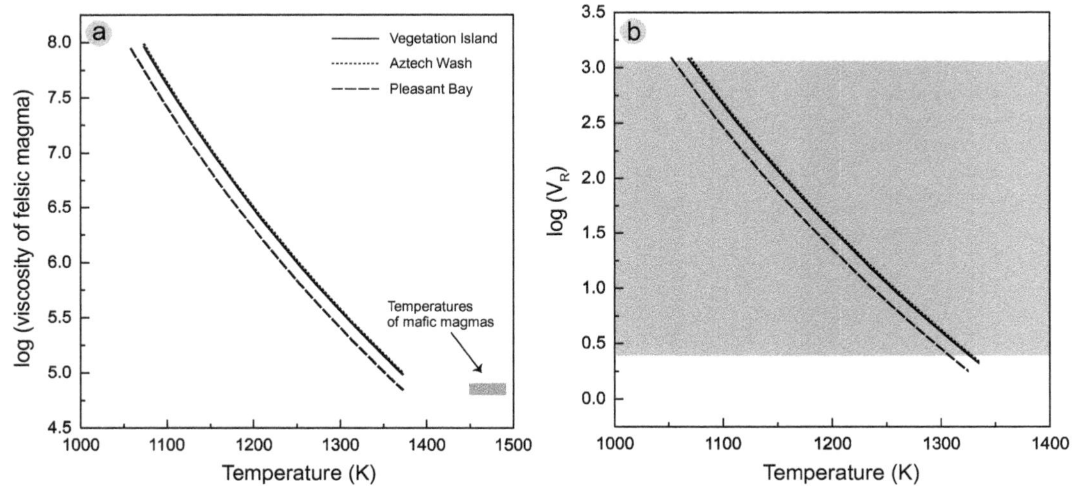

Fig. 6.10 Variation of the logarithm of viscosity of the felsic magma as a function of temperature. Gray bar represents the range of viscosity of mafic magmas in the three studied natural outcrops at $T = 1480$ k. Liquidus temperatures of mafic magmas were calculated at $P = 5000$ bars using the software MELTS (Ghiorso and Sack 1995); **b** variation of $\log(V_R)$ with temperature. The gray area represents viscosity ratios estimated for natural outcrops

decrease of viscosity of the latter due to the heat released by the mafic magma (e.g. Snyder and Tait 1998).

The injection of the mafic magma into the felsic one is considered as a progressive process occurring as hundreds of continuous injections (e.g. Snyder and Tait 1995; Bons et al. 2004). The continuous arrival of the mafic magma can provide, in principle, the heat needed to decrease the viscosity of the felsic magma. In addition, the volume ratio between the two magmas (mafic/felsic) rises abruptly, enhancing the increase in temperature of the felsic magma. This is supported by the fact that volume proportions of mafic and felsic magmas in the three case studies are of the order of 90% and 10%, respectively.

Such a conceptual model might explain the large variability of viscosity ratios estimated for the three natural outcrops considered here. In particular, it can be hypothesized that during the initial stages of injection of the mafic magma viscosity contrasts were high and this would have produced finger-like morphologies. As the influx of mafic magma proceeded in time, the felsic magma was progressively heated and viscosity contrasts decreased, leading to the formation of circular interfaces.

In order to test this idea we can consider a mafic magma at a temperature of ca. 1470 K with a viscosity of ca. $1.0 \cdot 10^5$ Pa s (e.g. Folch and Marti 1998; Jellinek and Kerr 1999). These values are suitable for mafic magmas with compositions analogous to the average composition of the mafic magmas in the three case studies (Perugini and Poli 2005, 2011). Following the idea suggested above, temperature and viscosity of the incoming mafic magma will remain almost constant, whereas temperature and viscosity of the felsic magma will change in time.

The viscosity of the felsic magma at different temperatures can be calculated using the empirical relationship derived by Hess and Dingwell (1996):

$$\ln(\eta) = [-3.545 + 0.833\ln(w)] + \frac{[9601 - 2368\ln(w)]}{\{T - [195.7 + 32.25\ln(w)]\}} \quad (6.2)$$

where η is viscosity (in Pa s), w is H_2O concentration (in wt.%), and T is temperature (in K). In the estimate of viscosity, the water content (w) of the felsic magma is considered to correspond to the average LOI [loss On Ignition] of natural samples (see Perugini and Poli 2005, 2011).

The graph of Fig. 6.10a shows the variation of viscosity of the felsic magmas for the three case studies plotted against temperature. The figure shows that as temperature increases, viscosities are progressively shifted towards the viscosity of the mafic magmas. The viscosity ratio (V_R) between magmas was calculated using data displayed in the plot of Fig. 6.10a. Figure 6.10b reports the variation of $\log(V_R)$ with temperature. For comparison, in the figure is also reported a gray-shaded area representing the range of $\log(V_R)$ estimated by fractal analysis on natural structures. The graph shows that there is complete overlap between $\log(V_R)$ values estimated by fractal analysis and those calculated. This indicates that the estimated V_R values can be considered reliable and in keeping with reasonable values of V_R that may occur in natural magmatic systems.

A further issue that is strictly connected to the previous one is the amount of mafic magmas needed to increase the temperature of felsic ones to achieve viscosities suitable to give V_R values comparable with those estimated. Following Folch and Marti (1998), the temperature variation in a felsic magma intruded by a mafic magma can be estimated as

$$\Delta T_f = \frac{\varphi \rho_m C_m}{\varphi \rho_m C_m + \rho_f C_f} (T_{mi} - T_{fi}) \qquad (6.3)$$

where T_{mi} and T_{fi} are the initial temperatures of the mafic and felsic magmas, respectively, C_m and C_f are their specific heat capacities, ρ_m and ρ_f are their densities, and φ is the volume fraction of injected magma. Parameters used in the calculation are as follows: $C_m = 1200$ JK^{-1}kg^{-1}, $C_f = 1000$ JK^{-1}kg^{-1}, $T_{mi} = 1490$ K, $T_{fi} = 1070$ K, $\rho_m = 2700$ kgm^{-3} and $\rho_f = 2300$ kgm^{-3} and are reasonable for magmas similar to those occurring in the considered outcrops (e.g. Folch and Marti 1998; Jellinek and Kerr 1999).

Calculation of φ reveals that, in order to encompass the range of temperature (from 1100 to 1330 K) of felsic magmas suitable to give the estimated V_R values, volume fractions of mafic magmas from ca. 0.05 to 0.90 are required. This indicates that the viscosity ratio V_R depends on the relative amount of mafic magma that is injected in the felsic host. In particular, injection of low volumes of mafic magma corresponds to low heat flux towards the felsic magma and, hence, a high viscosity contrast. Conversely, the arrival of larger volumes of mafic magma will cause a more vigorous heating of the felsic one, leading to a decrease of the viscosity ratio. This indicates that high fractions of mafic magmas are needed to reduce the viscosity of the felsic magma to the lowest values. This is in general agreement with field evidence where the relative volume of mafic magmas reaches values up to 90% of the entire outcrop. This large amount of high temperature mafic magma, therefore, can act as a powerful heat source, providing the necessary thermal energy for increasing the temperature of the felsic magma, up to 1330 K.

It is therefore clear that the most important factor acting to shift the viscosity of the felsic magma towards the viscosity of the mafic magma (i.e. towards low viscosity ratios) is due to thermal exchanges. Thermal diffusion strongly depends on the extension of contact interfaces; in detail, the efficiency of thermal exchanges increases as a function of the amount of contact surfaces. It appears plausible, hence, the idea that most of thermal exchanges occurred during the initial phases of the injection process, where large contact interfaces were available for thermal diffusion. Thermal exchanges play a key role in magma mixing systems because they can trigger convection dynamics. This can potentially cause the development of efficient and vigorous mixing processes throughout the whole magmatic mass, also driving extreme degrees of heterogeneity, which are commonly observed in igneous rocks. These aspects will be discussed in details in the next chapters.

References

Allen E, Boger DV (1988) The influence of rheological properties on mobility control in polymer-augmented waterflooding. Soc Petrol Eng (SPE). https://doi.org/10.2118/18097-ms

Bons PD et al (2004) Melt extraction and accumulation from partially molten rocks. Lithos 78(1–2):25–42. issn: 00244937. https://doi.org/10.1016/j.lithos.2004.04.041

Campbell IH, Turner JS (1986) The influence of viscosity on fountains in magma chambers. J Petrol 27(1):1–30

Cardoso SSS, Woods AW (1996) Interfacial turbulent mixing in stratified magma reservoirs. J Volcanol Geotherm Res 73(3–4):157–175

Cruden AR, Koyi F, Schmeling H (1995) Diapirical basal entrainment of mafic into felsic magma. Earth Planet Sci Lett 131:321–340

Di Vincenzo G, Rocchi S (1999) Origin and interaction of mafic and felsic magmas in an evolving late orogenic setting: the Early Paleozoic Terra Nova Intrusive Complex, Antarctica. Contrib Mineral Petrol 137:15–35

Falkner CM et al (1995) Petrogenesis and tectonic significance of the calcalkaline, bimodal Aztec Wash pluton, Eldorado Mountains, Colorado River extensional corridor. J Geophys Res Solid Earth 100(B6):10453–10476. issn: 01480227. https://doi.org/10.1029/94JB03357

Faulds JE, Geissman JW, Mawer CK (1990) Chapter 3: Structural development of a major extensional accommodation zone in the Basin and Range Province, northwestern Arizona and southern Nevada; Implications for kinematic models of continental extension. Mem Geol Soc Am 176(1):37–76. issn: 00721069. https://doi.org/10.1130/MEM176-p37

Folch A, Marti J (1998) The generation of overpressure in felsic magma chambers by replenishment. Earth Planet Sci Lett 163(1–4):301–314. issn: 0012821X. 00196–4. https://doi.org/10.1016/S0012-821X(98)

Ghiorso MS, Sack RO (1995) Chemical mass transfer in magmatic processes IV. Arevised and internally consistent thermodynamic model for the interpolation and extrapolation of liquid-solid equilibria in magmatic systems at elevated temperatures and pressures. Contrib Mineral Petrol 119(2–3):197–212. issn: 00107999. https://doi.org/10.1007/BF00307281

Hess KU, Dingwell DB (1996) Viscosities of hydrous leucogranitic melts: a non-Arrhenian model. Am Mineral 81(9–10):1297–1300. issn: 0003004X. https://doi.org/10.2138/am-1996-9-1031

Hogan J, Sinha A (1989) Compositional variation of plutonism in the coastal Maine magmatic province; mode of origin and tectonic setting. Maine Geological Survey Publications. https://digitalmaine.com/mgs_publications/79

Howard KA, John BE (1987) Crustal extension along a rooted system of imbricate low-angle faults: Colorado River extensional corridor, California and Arizona. Cont Ext Tecton Geol Soc [Lond] Spec Publ 28:299–311

Jellinek MA, Kerr RC (1999) Mixing and compositional layering produced by natural convection. Part 2. Applications to the differentiation of basaltic and silicic magma chambers, and komatiite lava flows. J Geophys Res 104:7203–7218

Mandelbrot BB (1982) The fractal geometry of nature. Freeman, W.H

Nittmann J, Daccord G, Eugene Stanley H (1985) Fractal growth viscous fingers: quantitative characterization of a fluid instability phenomenon. Nature 314(6007):141–144. https://doi.org/10.1038/314141a0

Perugini D, Poli G (2005) Viscous fingering during replenishment of felsic magma chambers by continuous inputs of mafic magmas: field evidence and fluid-mechanics experiments. Geology 33(1):5–8

Perugini D, Poli G (2011) Frontiers: intrusion of mafic magmas into felsic magma chambers: new insights from natural outcrops and fluidmechanics experiments. Ital J Geosci 130(1):3–15

Poli G, Tommasini S, Halliday AN (1996) Trace element and isotopic exchange during acid-basic magma interaction processes. Trans R Soc Edinb-Earth Sci 87(1–2):225–232. issn: 02635933. https://doi.org/10.1017/s0263593300006635

Saffman PG, Taylor G (1958) The penetration of a fluid into a porous medium or Hele-Shaw cell containing a more viscous liquid. Proc R Soc Lond. Ser A, Math Phys Sci 245(1242):312–329. issn: 00804630. https://doi.org/10.2307/100420

Snyder D, Tait S (1995) Replenishment of magma chambers: Comparison of fluid-mechanic experiments with field relations. Contrib Mineral Petrol 122(3):230–240

Snyder D, Tait S (1998) The imprint of basalt on the geochemistry of silicic magmas. Earth Planet Sci Lett 160(3–4):433–445

Sparks RSJ, Marshall LA (1986) Thermal and mechanical constraints on mixing between mafic and silicic magmas. J Volcanol Geotherm Res 29(1–4):99–124. issn: 03770273. https://doi.org/10.1016/0377-0273(86)90041-7

Sreenivasan KR et al (1989) The fractal geometry of interfaces and the multifractal distribution of dissipation in fully turbulent flows. Pure Appl Geophys 131(1–2):43–60

Waight TE et al (2001) Isotopic responses to basaltic injections into silicic magma chambers: a Whole-rock and microsampling study of macrorhythmic units in the Pleasant Bay layered gabbro-diorite complex, Maine, USA. Contrib Mineral Petrol 142(3):323–335. issn: 00107999. https://doi.org/10.1007/s004100100292

Wiebe RA (1993) The pleasant bay layered gabbro-diorite, coastal maine: ponding and crystallization of basaltic injections into a silicic magma chamber. J Petrol 34(3):461–489. issn: 00223530. https://doi.org/10.1093/petrology/34.3.461

The Development of Magma Mixing in Space and Time

7

When you can gather it up, start pulling it with your fingertips, allowing a spread of about 18 inches between your hands. Then fold it back on itself. Repeat the motion rhythmically. As the mass changes from a somewhat sticky, side-whiskered affair to a glistening crystal ribbon, start twisting, while folding and pulling.

Irma S. Rombauer and Marion Rombauer Becker, "Joy of cooking"

Abstract

In this chapter we will provide an explanation for the occurrence of magmatic enclaves and flow structures observed in the rocks. According to numerical simulations and physical experiments, the formation of enclaves and flow structures tightly reflects the dynamics occurring in the magmatic masses. They are, therefore, fundamental witnesses of the magma mixing process. However, the quantity and quality of information that they preserve is different in the plutonic and volcanic environment. The aim of this chapter is also to establish some general criteria and methods to extract the maximum information from the study of these structures in order to shed light into the petrological and geochemical evolution of magmatic systems that experienced magma mixing processes. We will also focus on the issue of the timing of magma mixing, illustrating some methods that can provide information on the timescales during which mixing processes can generate variably homogeneous magmatic masses.

7.1 Introduction

In this chapter we will provide an explanation for the occurrence of magmatic enclaves and flow structures observed in the rocks. As we have seen in Chap. 1, magmatic enclaves are one of the main indicators of magma mixing in both plutonic and volcanic rocks compared to flow structures that are, instead, mostly observed in volcanic rocks.

We will see that, according to numerical simulations and physical experiments presented in the previous chapters, the formation of enclaves and flow structures tightly reflects the dynamics occurring in the magmatic masses. They are, therefore, fundamental witnesses of the magma mixing process. However, the quantity and quality of information that they preserve is different in the plutonic and volcanic environment. The aim of this chapter is also to establish some general criteria and methods to extract the maximum information from the study of these structures in order to shed light into the petrological and geochemical evolution of magmatic systems that experienced magma mixing processes.

D. Perugini, *The Mixing of Magmas*, Advances in Volcanology,
https://doi.org/10.1007/978-3-030-81811-1_7

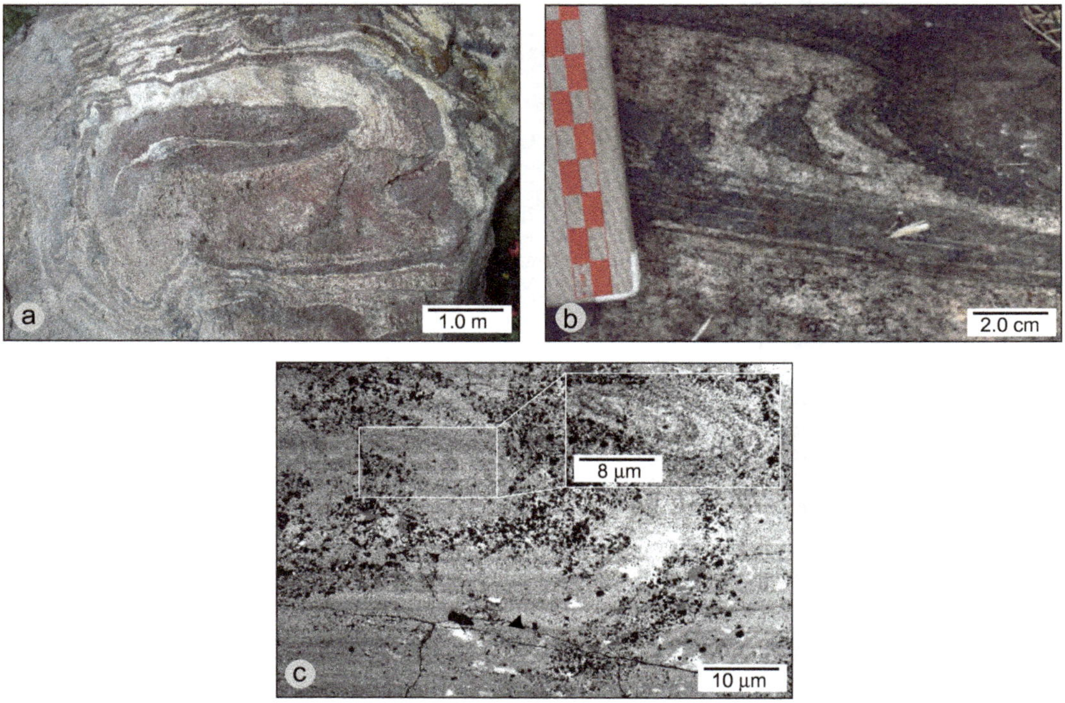

Fig. 7.1 Examples of magma mixing structures in a lava flow from the Island of Lesvos (Greece) showing the contemporaneous occurrence in the same system of flow structures and enclaves. These mixing patterns are repeated in the magmatic system at many length scales. Images in **c** are Back-Scattered Electron (BSE) pictures; inset shows a very small enclave

During our discussions we will also focus on the issue of the timing of magma mixing. In particular, we will illustrate some methods that can provide information on the timescales during which mixing processes can generate variably homogeneous magmatic masses.

7.2 Mafic Enclaves

7.2.1 Kinematic Significance of Magmatic Enclaves

Some representative pictures of magma mixing patterns in volcanic rocks are shown in Fig. 7.1. The three pictures belong to a lava flow cropping out on the Island of Lesvos (Greece). Enclaves and flow structures can be seen at several length scales. In the case of Fig. 7.1, we can observe them from the metre to the micrometre. Intuitively, therefore, there is a fractal cascade of mixing patterns in the rocks, in agreement with the chaotic nature of the process (Chap. 2).

In Chap. 4 we have discussed $2D$ and $3D$ numerical models that can be useful to follow in time the development of magma mixing processes. We have seen that in most cases the mixing system is constituted by two main dynamical regions where the mixing efficiency is extremely different. These were named "regular" (CR) and "chaotic regions" (AMR). In the light of those numerical simulations, we can infer that magmatic enclaves might represent blobs of the most mafic magma that did not experience strong stretching and folding dynamics. They had, therefore, the opportunity to survive the mixing process thanks to the close structure of flow fields in which they were trapped in during the time progression of mixing. This does not mean necessarily that they still represent the original mafic magma because, especially in

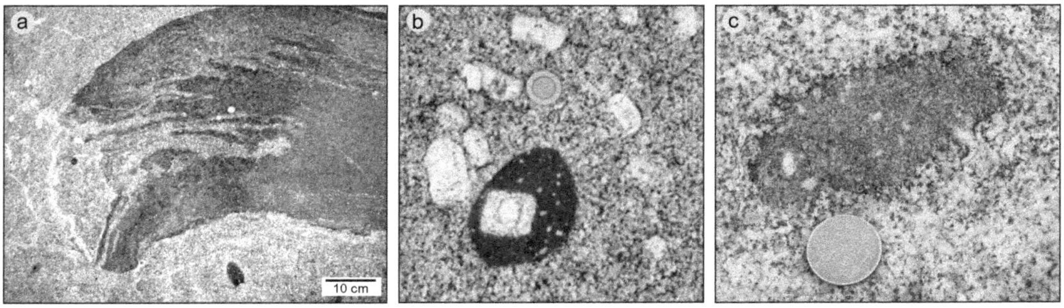

Fig. 7.2 Mafic enclaves from granitoid rocks showing the infiltration of the felsic magma (**a**) and the occurrence of minerals of the host magma (**b** and **c**); **a** and **b** Monte Capanne pluton (Italy); **c** pluton of Philippi (Greece)

the plutonic environment where the cooling of the magmatic mass requires longer times compared to volcanic rock, they might have experienced some interaction with the host felsic magma.

Figure 7.2 shows some mafic enclaves in granitoid rocks. Here we can observe that these enclaves experienced interaction with the host felsic melt. In particular, in Fig. 7.2a the felsic magma appears to invade the enclave on the left side, generating complex mixing patterns within the enclave. Figure 7.2b and c shows the occurrence of minerals (mainly feldspars) presumably belonging to the host magma in the mafic enclaves. Figure 7.2b is particularly enlightening because the k-feldspar megacrysts appear to mark the local flow field around the mafic enclave. One of these megacrysts was ultimately caught up by the enclave. Although these features can be observed in both volcanic and plutonic rocks, they are most frequently found in the latter. This indicates that, especially in plutonic rocks, enclaves can undergo internal mixing losing their original identity through effects of "dilution" from the host magma. We will discuss this issue in further details in the next sections.

The fact that enclaves can form within coherent regions does not imply that they necessarily will survive the mixing process. In fact, numerical models and high-temperature experiments (Chaps. 4 and 5) have shown that, if the flow field is changed in time (for example changing the control parameter k of the sine flow map) enclaves cannot be preserved. This happens because the coherent regions present in the system are destroyed by the newly formed flow field. In these conditions, enclaves might "fall" into the chaotic regions and, here, due to the efficient stretching and folding dynamics they can be dispersed in the host magma. In these circumstances we may be unable to recognize macroscopic indications of magma mixing and, at first sight, the magmatic mass might appear compositionally homogeneous. This is probably the worse context for a petrologist because it represents a condition of undecidability: is the magmatic mass homogeneous because it is constituted by a single magma or it is the result of perfect hybridization between two end-member magmas? Answering such a question is not easy. However, as we will see in the next chapter, mineral phases showing physical and chemical disequilibrium can help us to partially address the problem.

7.2.2 Morphological Analysis of Magmatic Enclaves in the Volcanic Environment

Magmatic enclaves can display a variety of morphologies and their analysis can provide insights on to the nature of the mixing process. The morphological analysis of enclaves is best performed in volcanic rocks because (i) the contact with the host magma is generally sharp and (ii) their morphology is likely to be less influenced by secondary processes and more closely related to the pristine shape.

Figure 7.3 shows some mafic (latitic) enclaves in the A.D. 1793 Pietre Cotte rhyolitic lava flow

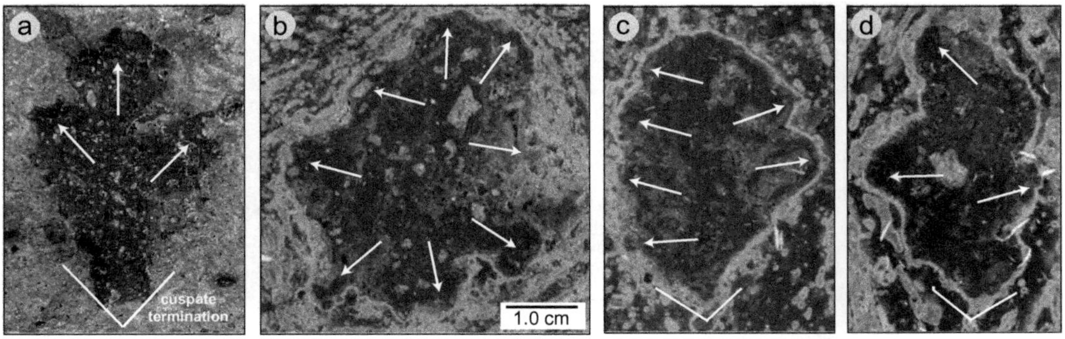

Fig. 7.3 Mafic enclaves from the Pietre cotte lava flow (Island of Vulcano, Italy). Note the enclave morphology where fingers of the mafic magma tend to penetrate the felsic host

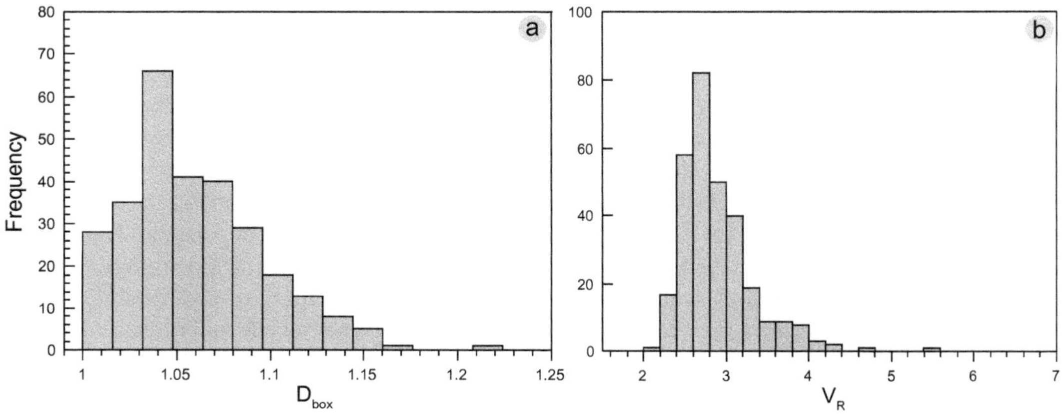

Fig. 7.4 **a** Frequency histograms of measured values of D_{box} and **b** calculated viscosity ratios V_R for enclaves occurring in the Pietre Cotte lava flow

(Island of Vulcano, Aeolian Islands, Italy). Mafic enclaves amount to 5–6% of the entire outcrop. In general, they have irregular margins and the contact interface with the rhyolitic host is well marked. In spite of their morphological variability, enclaves display some common features: (i) the contact interface with the rhyolite shows fingers pushing towards the host magma and (ii) in several samples enclaves have sharp and cuspate terminations (Fig. 7.3).

Enclave morphology can be measured by estimating their fractal dimension (D_{box}). Also in this case the box-counting technique (see Chap. 2) can be applied.

The frequency histogram in Fig. 7.4a shows the measured variability of D_{box} values and indicates that the fractal dimension of latitic enclaves ranges from 1.00 to 1.21, with a modal value of 1.04.

Previous works on the Pietre Cotte lava flow indicated that the morphology of enclaves was inherited from the magmatic reservoir where the mafic magma entered in contact with the felsic one (Perugini et al. 2007; Piochi et al. 2009). As we have seen in Chap. 6, the injection of mafic magmas into felsic hosts triggers instability of the interfaces between the two magmas. This produces the development of fingers of the mafic magma pushing and penetrating the felsic one.

The shape of viscous fingering patterns is a function of viscosity ratio (V_R). The larger V_R, the larger is the irregularity of the pattern. If after the injection process the system remains in static conditions (e.g. convection dynamics do not develop) the fingers might be steady. On the contrary, if convection dynamics develop (e.g. Snyder and Tait 1996; Folch and Marti 1998; Snyder 2000),

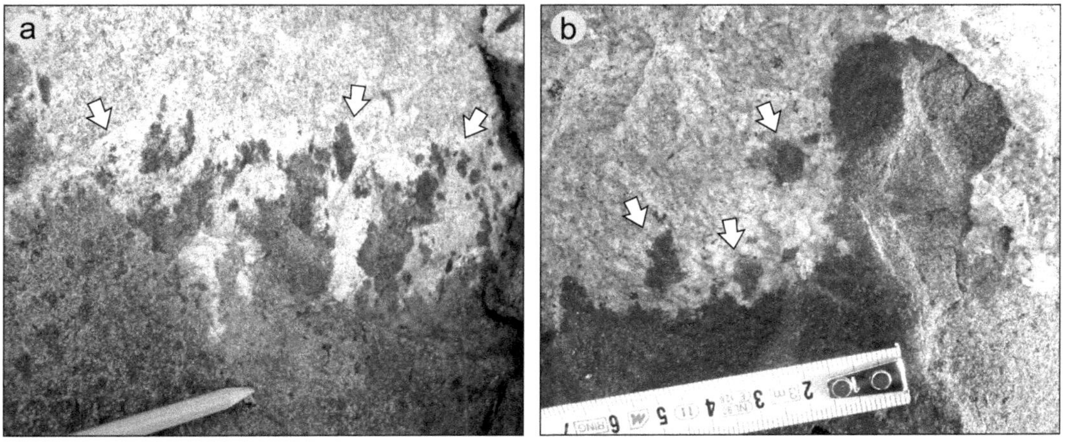

Fig. 7.5 Viscous fingering structures occurring in the Vegetation Island outcrops (Terra Nova Intrusive Complex, Antarctica; Perugini and Poli 2005). Note the widespread presence of mafic enclaves in the felsic mass (indicated by white arrows) of variable size resulting from the disruption of viscous fingering patterns

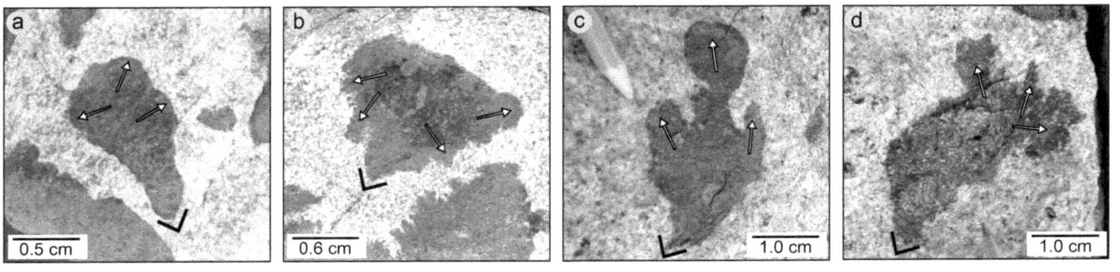

Fig. 7.6 Mafic enclaves generated by the fragmentation of viscous fingering structures at Vegetation Island outcrops (Terra Nova Intrusive Complex, Antarctica). White arrows indicate fingers of the mafic magma propagating towards the felsic host; the occurrence of cuspate terminations is also indicated

fragmentation of the viscous fingering structures will occur. This processes will be mostly experienced by the fingers, that are the most exposed to fragmentation.

These effects can be observed in viscous fingering structures from Vegetation Island outcrops (Terra Nova Intrusive Complex, Antarctica; Fig. 7.5) where several fragments detached from fingers of the mafic magma and are incorporated as enclaves in the felsic host. Figure 7.6 shows detailed pictures of enclaves detached from natural viscous fingering patterns from the Terra Nova Intrusive Complex outcrop. Observation of these pictures indicates a strong morphological similarity between enclaves occurring in the lava flow (Fig. 7.3) and those generated by the disruption of the viscous fingering patterns (Fig. 7.6). In particular, in both cases, enclaves display irregular morphologies represented by fingers of the mafic magma propagating towards the felsic host. Additionally, comparison between enclave morphology in Figs. 7.3 and 7.6 highlights the occurrence of cuspate terminations in the enclaves from the lava flow and in those generated by the disruption of viscous fingering structures. Cuspate edges may represent the points of detachment of the enclaves from the mafic mass during the fragmentation of the fingers. This morphological similarity argues in favor of the hypothesis that the occurrence of mafic enclaves in the lava flow could be associated with magma chamber dynamics.

Comparative analysis between enclaves in the lava flow and those generated by the disruption of viscous fingering structures in plutonic rocks can be performed studying the fragment size distribu-

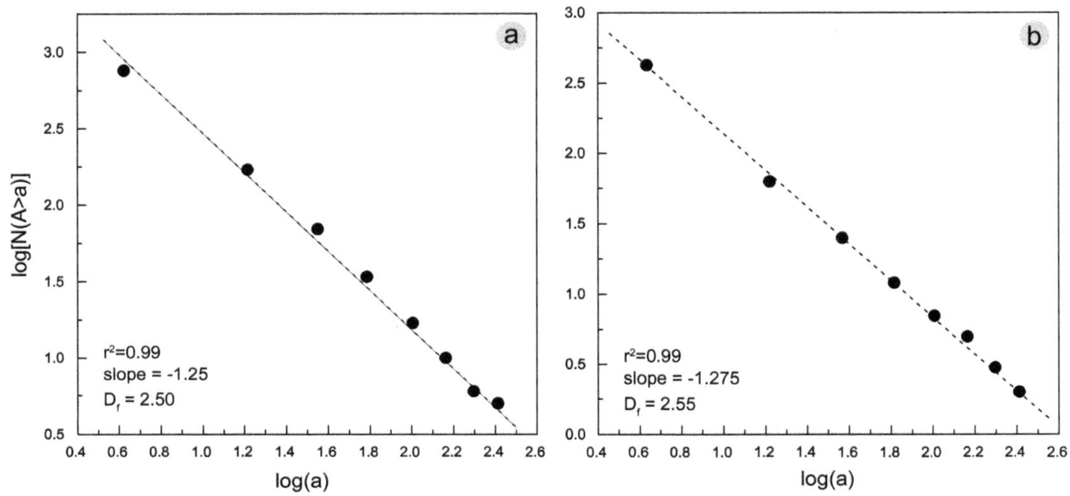

Fig. 7.7 Variation of the logarithm of cumulative number of mafic enclaves with areas A larger than comparative area a ($\log[N(A > a)]$) against $\log(a)$, for enclaves from the Pietre Cotte lava flow (**a**) and enclaves formed by disruption of viscous fingering structures (**b**)

tion of enclaves in both geological contexts. This can be done using the fractal fragmentation approach that we have illustrated in Chap. 2. In that case we have used the linear size of fragments produced by the fragmentation model (Fig. 2.12). As here we are dealing with more complex morphologies, we can apply fractal statistics considering the surface area of enclaves. In this case Eq. 2.8 becomes (Mandelbrot 1982)

$$N(A > a) = ka^{-D_f/2} \tag{7.1}$$

with

$$D_f = -2m \tag{7.2}$$

Figure 7.7 shows the variation of the logarithm of cumulative number of mafic enclaves with areas A larger than comparative area a ($\log[N(A > a)]$) against $\log(a)$, for enclaves from the lava flow (Fig. 7.7a) and those formed by disruption of viscous fingering structures (Fig. 7.7b). In both cases, the graph shows that data points are disposed along a straight line, fulfilling the requirement for a fractal distribution. Linear interpolation of data gives a slope (m) equal to -1.25 and -1.275 for enclaves from the lava flow and granitoid rocks, respectively. According to Eq. 7.2, this corresponds to similar values of fragmentation fractal dimension $D_f = 2.50$ and $D_f = 2.55$, respectively. This can be considered an indication that enclaves from both geological contexts were generated by the similar dynamics.

Following this idea and given that the fractal dimension (D_{box}) of enclaves reflects the viscosity ratio (V_R) between magmas, the values of V_R was estimated using the empirical relationship reported in Chap. 6 (Eq. 6.1). The histogram in Fig. 7.4b displays the variability of viscosity ratios estimated for the enclaves in the Pietre Cotte lava flow. Results show that (V_R) ranges from ca. $V_R = 2.3$ to $V_R = 5.5$ with most values clustered around $V_R = 2.6$ [$\log(V_R) = 0.42$]. It is important to recall that this value is very similar to the modal values of V_R we have estimated for the three case studies of viscous fingering in Chap. 6 ($\log(V_R)$ ca. 0.5–0.6), confirming that the intuition that enclaves from the lava flow represent remnants of viscous fingering structure can be considered reasonable.

7.2.3 Dilution of Mafic Enclaves by Diffusion and Infiltration of the Host Magma

As mentioned before, mafic enclaves in granitoid rocks are likely to have experienced interaction with the host felsic magma. This idea was essentially based on the observation of pictures shown in Fig. 7.2. We will now attempt to quantify this process studying mafic microgranular enclaves (MME) occurring in the granodioritic bodies of the Sithonia Plutonic Complex (Northern Greece; Perugini et al. 2003; Christofides et al. 2007).

These enclaves can be divided into two groups based on their geochemical composition: monzodioritic and quartz-monzodioritic (hereafter referred to as Mz-MME, QMz-MME). The two groups of enclaves have different compositions and follow a common trend with their hosts. Mz-MME are produced by lower degrees of interaction; QMz-MME lie farther from the mafic end-member and are the result of larger degrees of interaction (Perugini et al. 2003).

Observation of the contacts between the enclaves and the felsic host indicates that they may have suffered infiltration of the felsic magma and, consequently, their composition might have changed by this process (Fig. 7.8). Figure 7.8 shows that enclave boundaries are not sharp and minerals from the host rock are now incorporated into the enclaves. As we have suggested above, MME are likely to represent portions of the mafic magma that were trapped within regular islands. These are elliptic sets and crystals present in the host magma travel around them along tangential flow lines. Crystals that are transported along flow lines, close to enclaves, may be swallowed up into the enclaves. However, this depends on their ability to align along these lines. A key parameter controlling this process is the crystal shape. We have seen before that feldspars are commonly found into the enclaves. Biotite is less common (Fig. 7.8). This may be a consequence of the fact that the prismatic habit of plagioclase does not allow the mineral to align itself perfectly along the flow lines and consequently it can be easily caught up into the enclaves. Biotite, on the other hand, characterized by a tabular habit, can adapt well to the direction of the flow lines and hence remain in the host magma mantling the enclaves. The fact that plagioclase crystals penetrate within enclaves implies that also the liquid from the host magma can penetrate into the enclaves inducing dilution of the mafic magma.

To test quantitatively whether enclaves suffered the effects of dilution from the host felsic magma X-ray maps of major elements (Fig. 7.9) were collected on MME thin sections (Perugini and Poli 2004). X-ray maps can be used to determine the degree of compositional disorder of MME, which is expected to correlate with their degree of dilution from the host magma and, hence, ultimately with their geochemical evolution.

The estimate of the compositional disorder of enclaves can be performed measuring the Shannon entropy (S) of the $2D$ X-ray maps. Note that this method is identical to that used to determine the Shannon entropy on compositional transects in Chap. 5, but in $2D$. We divide the map into a grid constituted by a number M of cells c_i with area a_i, such that $\sum_{i=1}^{M} a_i$, the total area of the map (Fig. 7.9c). As reported in Baranger et al. (2002), S is estimated as

$$S(t) = -\sum_{i=1}^{M} p_i(t) \log p_i(t) \tag{7.3}$$

where $p_i(t)$ is the probability that a given chemical element is situated into cell c_i at time t. $p_i(t)$ is calculated for each cell as the ratio between the number of points in the cell (N_c) and the total number of points in the map (N_{tot}). $S(t)$ is the sum of the values of $p_i(t) \log p_i(t)$ calculated considering all cells M. As discussed before, S increases with time.

Figure 7.10 reports the measured values of S for representative elements as a function of the CaO content of enclaves. CaO is an element that can track the geochemical evolution of MME (Perugini et al. 2003). Since the mixing intensity increases with the passing of time, decreasing amounts of CaO in the MME correspond to increasing mixing times. The graphs show that S values increase with the intensity of geochemical interaction of the enclaves (i.e. towards lower CaO contents). However, the two different groups of enclaves (Mz-

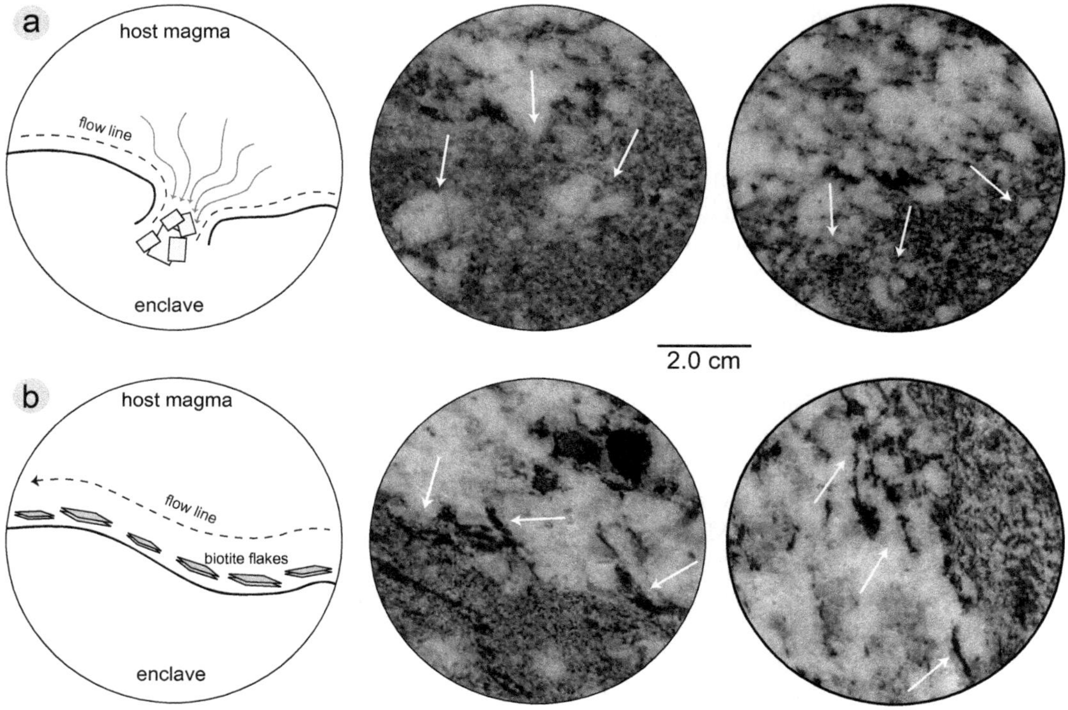

Fig. 7.8 **a** Schematic drawing and pictures of natural rocks showing the transfer of feldspars and liquid to the enclave magma; feldspar xenocrysts are indicated with white arrows; **b** Schematic drawing and pictures of natural rocks showing the orientation of biotite crystals along the contact between the host magma and the enclave; biotite crystals are indicated with white arrows

MME and QMz-MME) display different rates of increase of compositional disorder (*S*), as reflected by the different slopes of the linear regression for each enclave group. In particular, Mz-MME show higher slopes relative to QMz-MME. A further feature which we can observe in Fig. 7.10 is that the increase of the degree of compositional disorder (*S*) from Mz-MME to QMz-MME occurs at different rates (i.e. different slopes of the linear fitting in Fig. 7.10) for the different chemical elements. For example, the increase of *S* in Mz-MME for Mg (slope = 0.18) is faster, at the same extent of magma interaction (i.e. same CaO content), than the increase of *S* for K (slope = 0.09). A possible reason for this behaviour can be found in the relative concentrations of chemical elements in the end-member magmas. In the case of elements having the same concentration in the end-members, the mixing process would not change the concentration of those elements. As a consequence, the variation of *S* will be low. If, on the other hand, elements were present in different concentrations, we can expect to observe a much larger variation in *S*.

MME and host rocks have both experienced mixing processes. Therefore, reconstructing the original compositions of the mafic and felsic end-members is not easy. Nevertheless, it appears reasonable to hypothesize that the compositions of the end-member mafic and felsic magma were closer to the composition of the least evolved MME (Mz-MME) and the most differentiated host rocks, respectively (Perugini et al. 2003). For example, the abundances of MgO and K_2O in the most mafic enclave and in the most felsic host, indicate that the end-members presumably had different MgO concentrations (most mafic MME ca. 4.0 wt.%, most felsic host rocks ca. 0.1 wt.%), but similar K_2O concentrations (most mafic MME 3.77 wt.%, most felsic host rocks 3.79 wt.%).

Nevertheless, the puzzling issue of what processes acted to generate the relationship between

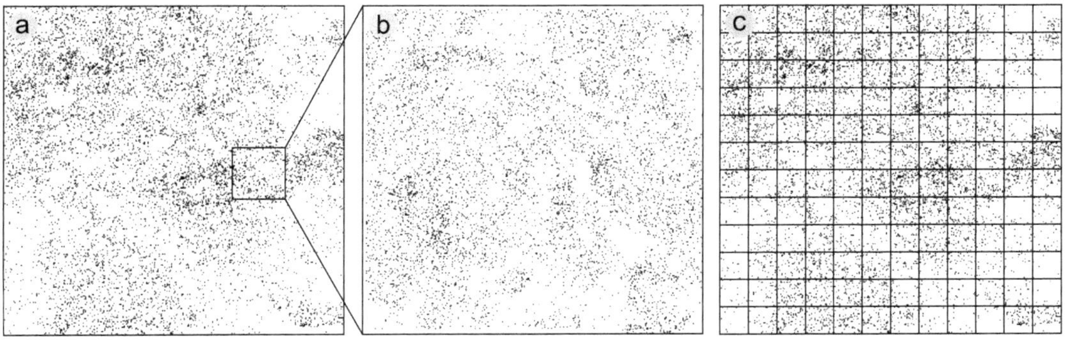

Fig. 7.9 **a** X-ray elemental map displaying the distribution of Ca in one of the analyzed MME; **b** enlargement of part of the map displayed in **a**; **c** example of grid used for the calculation of the degree of compositional disorder (S) of MME

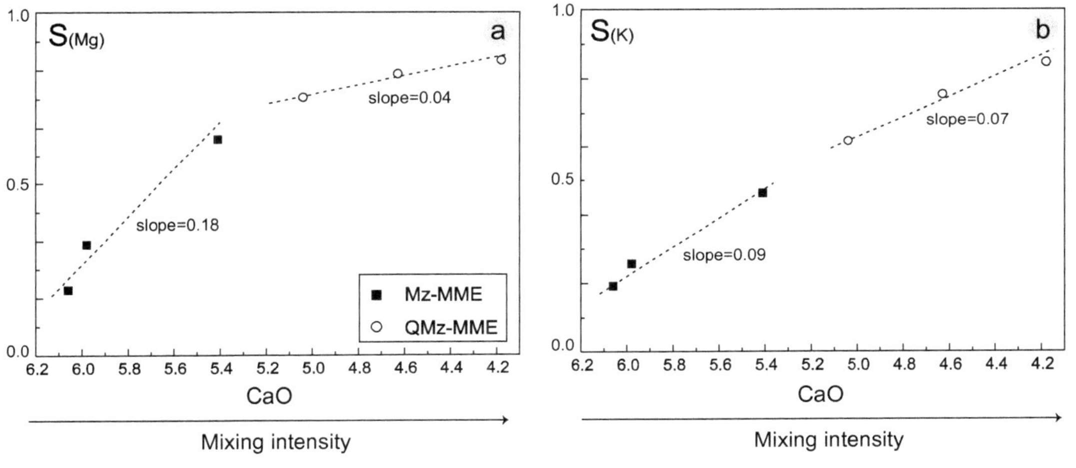

Fig. 7.10 Variation of the degree of compositional disorder (S) as a function of the degree of geochemical interaction (mixing intensity) of MME

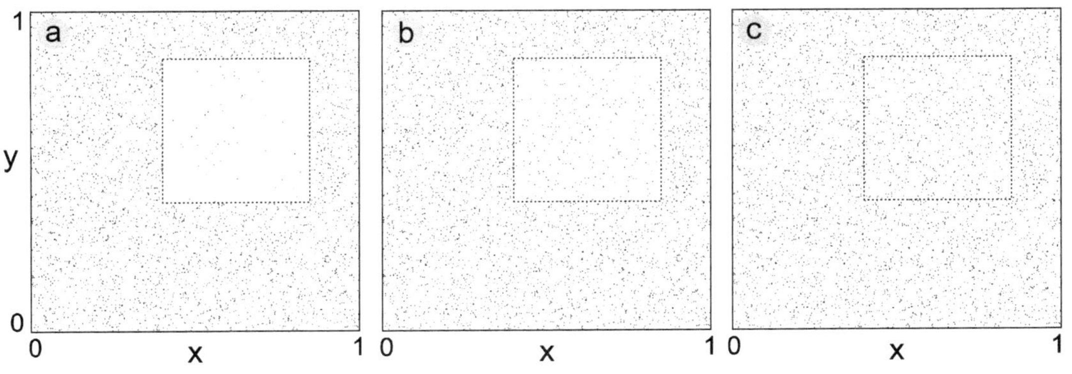

Fig. 7.11 Initial configuration of mixing systems in which the felsic and mafic magmas have different abundances of the chemical element X. In the felsic magma (dotted square) concentration of X increases from **a** to **c**; in the mafic magma concentration of X is constant

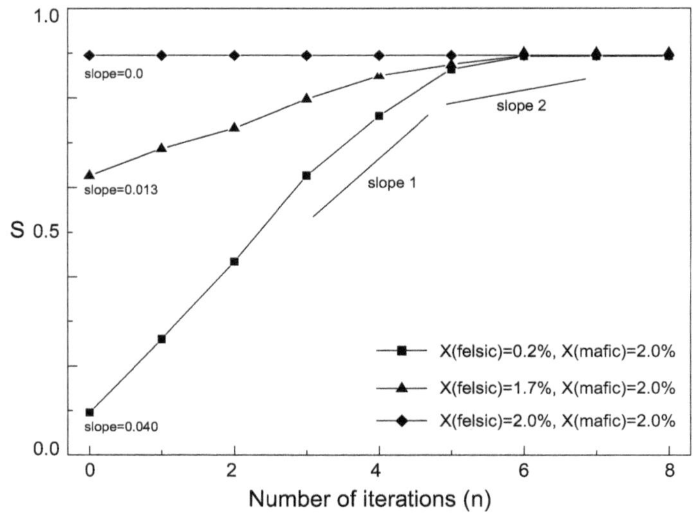

Fig. 7.12 Variation of S with mixing time for the systems displayed in Fig. 7.11. The slope of the variation of S versus n decreases as the difference in concentration of element X in the felsic and mafic magma decreases

the extent of geochemical interaction in the MME and their degree of compositional disorder (S) remains. In the attempt to provide an explanation, we can assume that S is linked to the distribution of chemical elements within the enclaves. Given that enclaves are completely crystallized, element distributions have to be related to the disposition of minerals in the enclaves. This, in turn, has to be a function of the composition of the magma forming MME. Stated this, S values must depend on the level of homogeneity of chemical elements in the MME.

We have seen before that it is possible that enclaves underwent dilution from the host magma. This agrees with the results of Grasset and Albarede (1994), who showed that magmatic enclaves can suffer increasing compositional exchanges with the host magma, generating more differentiated enclaves with the progression of time.

To test whether different initial abundances of chemical elements can explain the different rates of S variation, we can perform a simple numerical experiment of enclave dilution. We consider a mafic magma, represented by a square and containing an element (X), with a given concentration (black dots; Fig. 7.11). Within the mafic magma, we place a square portion of felsic magma in which the same element (X) is present with a different concentration (Fig. 7.11). This system is mixed using the Arnold's cat map presented in Chap. 2 to redistributed the element in the mafic magma. The variation of compositional disorder (S) is calculated for different systems with variable concentrations of the element X in the mafic and felsic magma.

Figure 7.12 shows the variation of compositional disorder (S) for each element as the number of iterations of the dynamical system increases. The graph shows that, for all systems, S increases linearly during the first stages of the dilution process. Successively, it saturates towards a maximum constant value. The plot also shows that when the abundances of element X are very different in the felsic and mafic magma [for example $X(\text{felsic}) = 0.2\%$ and $X(\text{mafic}) = 2.0\%$], S increases very quickly (high slope values). On the contrary, if element concentrations are similar in both magmas [for example $X(\text{felsic}) = 1.7\%$ and $X(\text{mafic}) = 2.0\%$] the increase of S is slower.

Comparing the variation of S in the simulations (Fig. 7.12) and in the MME (Fig. 7.10), we can observe that: (i) when the initial concentrations of chemical elements are similar in the two magmas (e.g. K_2O), S increases slowly (low slope) and (ii) when the concentrations of elements are different (e.g. MgO), S increases faster (larger slope).

Therefore, as a first approximation, it appears that the idea of dilution of the mafic enclaves by the felsic magma can explain the observed re-

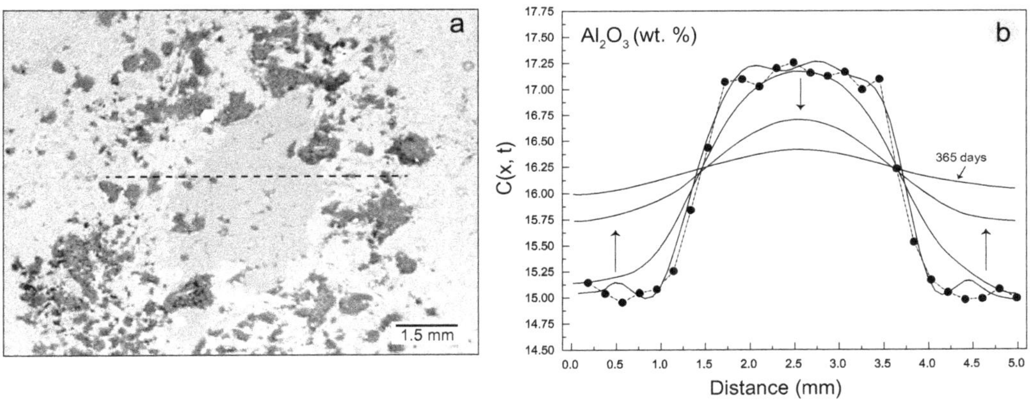

Fig. 7.13 **a** Back scattered electron image of a trachytic enclave in the rhyolitic magma from the Lesvos lava flow. The dashed line indicates the path followed by EMPA analysis; **b** Fourier series fitting of the variation of Al_2O_3 across the enclave. Black dots represent the measured compositional variability. The variation of $C(x, t)$ against x at different times is also shown. Modified from Perugini et al. (2003)

lationship between the degree of compositional heterogeneity of enclaves and their geochemical characteristics. Given that the dilution process is considered to be triggered by the disruption of the mafic-felsic interface from mineral phases, we can regard crystals (in this case feldspars) as trojan horses that allowed the felsic magma to infiltrate and mix with the enclaves.

7.2.4 Timing of Homogenization of Mafic Enclaves

At this stage of our study of magmatic enclaves it might be worth attempting to address a cardinal question: assuming that enclaves formed because the mafic magma was trapped within the coherent regions, how long would it take for chemical diffusion to homogenize the enclave and erase any trace of it from the magmatic system?

Figure 7.13a shows an enclave of trachytic composition occurring in a rhyolitic lava flow cropping out on the Island of Lesvos (Greece; Perugini et al. 2003). The dashed line in the figure represents the path followed by EMPA to determine major element concentrations in the enclave and the host magma. Note that the enclave is ca. 3–5 mm in diameter (Fig. 7.13a). The black dots in Fig. 7.13b show the variation of Al_2O_3. It defines a continuous bell-shaped pattern passing from the felsic magma to the enclave. A plateau of constant values occurs at the center of the enclave. These patterns are typically encountered during the study of chemical diffusion (e.g. Crank 1975; Kuo et al. 1997). This indicates that diffusive exchanges occurred between the trachytic and rhyolitic melt. The highest and lowest values of Al_2O_3 can be assumed as those in the initial end-members.

We can use the compositional variability in Fig. 7.13b to estimate time required to homogenize the enclave with the host magma assuming that the magmatic system remained in the molten state. This time can be estimated using the numerical solution of the diffusion equation for an infinite medium with an initial concentration $C_0(x, t)$ being a periodic function in space (Albarede 1995):

$$C(x, t) = C_0(x) \exp\left(-\frac{4\pi Dt}{\lambda^2}\right) \tag{7.4}$$

where x is space, t is time, D is the diffusion coefficient and λ is the wavelength. The variation of concentration through the enclave is fitted using the Fourier series technique whose polynomial expansion is used as the periodic function in the simulation of the diffusion process.

Homogenization requires that no traces of the original structure can be detected. This implies that the homogenization time has to be calculated for those chemical elements having the lowest diffusivities. Among them Al_2O_3 is one of the best candidates (diffusion coefficient of the order of 10^{-10} cm^2/s; e.g. Baker 1990) for magmas similar to those considered here.

Figure 7.13b shows the fitting of data by Fourier series expansion along with the measured values. The variation of concentration $C(x, t)$ against distance (x) with the passing of time is also shown. We can observe that the variation of concentration $C(x, t)$ is progressively smoothed with time. We can consider that the system is homogeneous when the variation of concentration falls within the limits of measurements by EMPA. Results of the modeling indicate that the time required for the complete homogenization of the enclave is of the order of one year. This timescale might appear quite short compared to typical lifetimes of magmatic reservoirs. However, we need to consider that the enclave is very small in size, of the order of a few millimeters. Increasing the enclave size would dramatically extend the period of time required for homogenization. As an example, if we consider an enclave with a diameter of one meter, the homogenization time increases to $10^5 - 10^6$ years (Petrelli et al. 2006).

It is worth comparing these timescales with the typical lifetimes of magma chambers. Heat flux calculations for magma chambers of a few kilometers suggest thermal lifetimes of about 10^4 years (e.g. Furlong and Myers 1985). Longer lifetimes are also possible (e.g. Volpe and Hammond 1991; Davies et al. 1994; Heath et al. 1988), but we prefer to be conservative in order to define the lower limits for the processes to operate efficiently. Therefore, assuming that a period of 10^4 years is a possible lower limit for the thermal lifetime of a magma chamber, only enclaves with diameters smaller than twelve centimeters can be completely homogenized.

7.3 Flow Structures

7.3.1 Kinematic Significance of Flow Structures

Flow structures preserved in volcanic rocks represent snapshots of the mixing process which has been frozen in time by an eruption. They can coexist with magmatic enclaves and propagate in the magmatic mass from the meter to the micrometer length scale (Fig. 7.1). Some examples of flow structures in volcanic rocks are shown in Fig. 7.14. These patterns mark the flow fields that developed during mixing and allow us to visualize the outcome of the stretching and folding process. As we have seen in Chap. 3, flow structures can form in the magmatic masses because of the occurrence of hyperbolic sets, resulting in the development of the chaotic regions in which stretching and folding are very efficient.

Although generally we are limited to $2D$ observations (i.e. $2D$ images) of mixing patterns in the rocks, in some cases we can add one more dimension and attempt to visualize mixing patterns in $3D$. This can be done using a variety of $3D$ reconstructions including X-ray micro-CT or serial slicing/lapping of samples. $3D$ observations allow us to perceive better the space distribution of flow structure and provide a powerful tool to penetrate in greater detail into the complexity of the mixing process.

Figure 7.15 shows the $3D$ reconstruction of a representative mixed pumice sample from the pyroclastic deposits of the Pollara eruption (Island of Salina, Aeolian Islands, Italy). Here eruptive products are characterized by banded pumices which are the result of mixing between an andesitic and a rhyolitic magma (e.g. De Rosa et al. 2002). A number of samples was reconstructed in $3D$ by serial lapping and digital montage (Perugini et al. 2008).

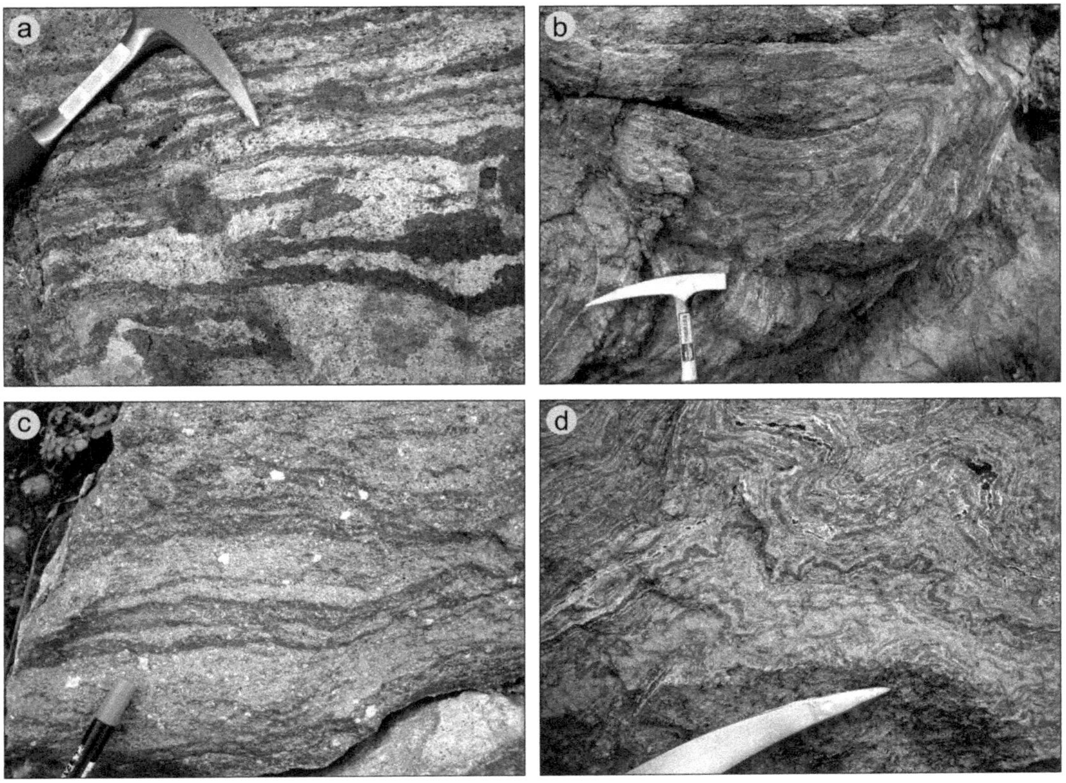

Fig. 7.14 Examples of flow structures from a lava flow of Lesvos (Greece) showing complex mixing patterns formed by the development of stretching and folding dynamics during magma mixing

An important result obtained by De Rosa et al. (2002) for Pollara rocks is that the SiO_2 content of the samples correlate with the grey value of sample images. This relationship is shown in the plot of Fig. 7.16 and allows us to use outcrop pictures to perform both digital image and compositional analysis based on the gray shades of the samples.

Figure 7.17a shows one of the samples in which we can observe that what we regarded as flow structures are actually surfaces of concentration spreading and overlapping in the 3D space. Using the relationship shown in Fig. 7.16, compositional histograms of the variability of SiO_2 concentration throughout the 3D sample can be constructed. The compositional histogram for the sample is shown Fig. 7.17b. The histogram comprises all compositions from the andesite to the rhyolite and defines a bell-shape pattern which is skewed on the right side, towards the rhyolitic and-member. In Chap. 4 we have discussed the meaning of compositional histograms in mixing systems. In particular, the maximum frequency of the histogram corresponds to the hybrid concentration that the system would attain if the mixing process was completed. In the case of the sample shown in the Fig. 7.17, it corresponds to a SiO_2 concentration of 74.10 wt.%. The composition of the hybrid can be used to segment the reconstructed 3D volume to visualize sample areas in which the hybrid concentration was generated by the mixing process. Figure 7.17c shows the distribution of the hybrid concentration in the sample and indicates that the formation of hybrid volumes is tightly connected with the spatial distribution of flow structures, following the same spatial orientation. The large surface area associated with flow structures, hence, favors the formation of hybrid compositions. This is the reason why in the chaotic regions the formation of homogeneous compositional domains is strongly enhanced compared to regular islands.

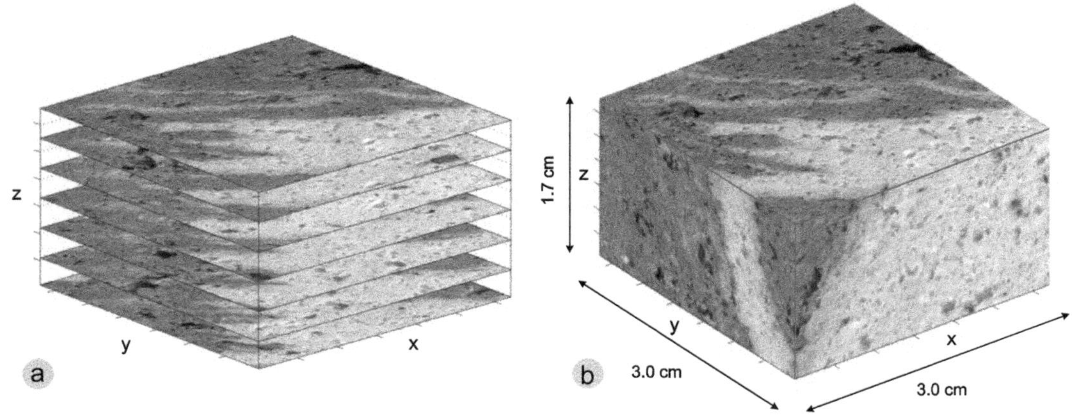

Fig. 7.15 Example of $3D$ reconstruction of a rock sample from the pyroclastic deposit of Pollara (Island of Salina, Italy); **a** montage of digital images (slices) obtained by serial lapping of the pumice sample; **b** $3D$ reconstructed sample. After Perugini et al. (2008)

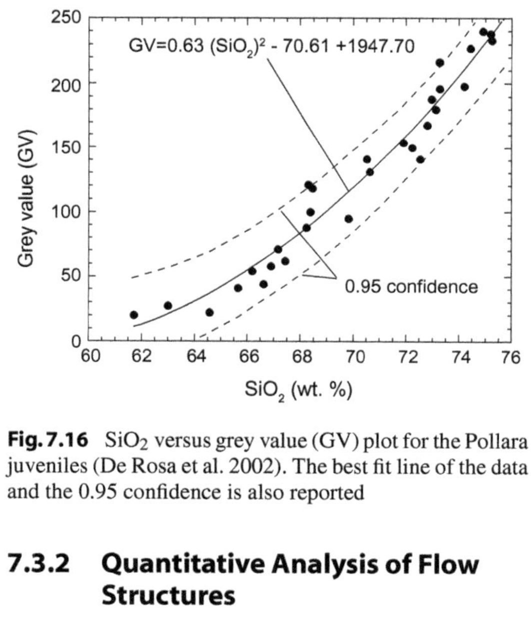

Fig. 7.16 SiO_2 versus grey value (GV) plot for the Pollara juveniles (De Rosa et al. 2002). The best fit line of the data and the 0.95 confidence is also reported

7.3.2 Quantitative Analysis of Flow Structures

Flow structures in volcanic rocks can be used to quantify the mixing process using image analysis techniques. This can be helpful, in combination with numerical models, to obtain further insights on the extent of mixing preserved in the rocks.

Given the fractal nature of flow structures (Fig. 7.1), their quantitative analysis can be performed using fractal geometry methods. In particular, as discussed in Chap. 2, one of the most useful techniques for fractal analysis of digital images is the box-counting method. Application of this method requires digital images to be segmented in order to render the flow structure and the background of the images in the black and white color, respectively. Figure 7.18 shows this process applied to a representative flow structure preserved in a lava flow on the Island of Lesvos. Black and white pictures have been used to measure their fractal dimension (D_{box}; see Chap. 2 for details). Along with D_{box}, also the length of interfaces (interfacial perimeter, IP, in pixels) between the two magmas was measured.

Measurements were performed on flow structures from three different lava flows from the Island of Lesvos (Greece), Vulcano and Salina (Aeolian Islands, Italy) (Perugini et al. 2003). The plot in Fig. 7.19a shows the variation of D_{box} as a function of $\log(IP)$ for the three lava flows. In general, the plot shows a linear relationship between these two variables for the three data sets. In addition, the three groups of flow structures display different slopes of the linear fitting.

The plot of Fig. 7.19b reports the variation of D_{box} versus $\log(IP)$ for numerical simulations performed using the sine flow dynamical system (Chap. 4) considering only advection (i.e. measurements were performed on mixing patterns simulated without considering chemical

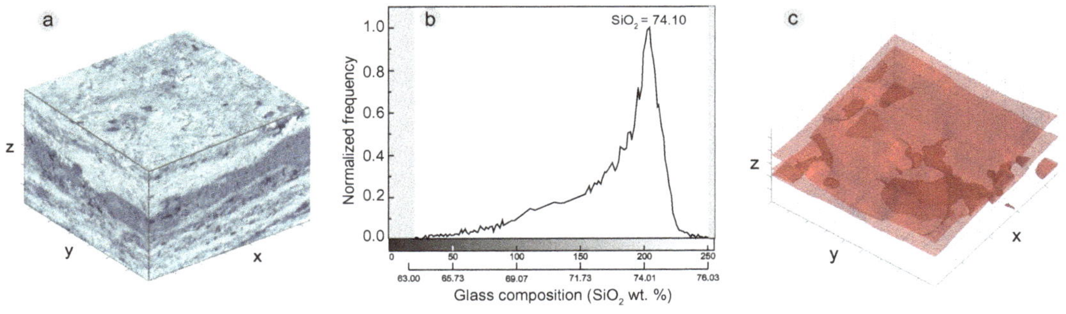

Fig. 7.17 Example of extraction of the hybrid composition from a reconstructed $3D$ sample from the Pollara pyroclastic deposit: **a** $3D$ sample; **b** compositional histogram of the sample; **c** $3D$ distribution of hybrid composition in the sample

Fig. 7.18 **a** Example of a magma mixing structure from a lava flow cropping out on the Island of Lesvos; **b** thresholded structure on which the length of interfaces (IP) and fractal dimension (D_{box}) have been measured

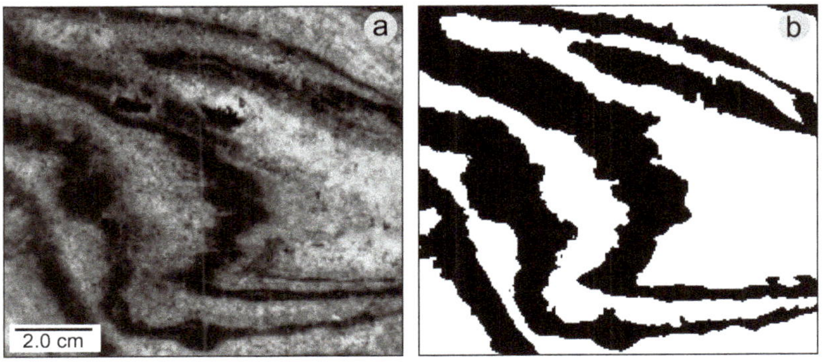

diffusion). The graph shows that, analogous to the natural structures, there is a linear increase of D_{box} with $\log(IP)$. Noteworthy is the fact that the structures produced numerically show increasing slopes of the linear fitting as the values of the control parameter k increases (i.e. increasing mixing efficiency).

On the basis of the results presented in Fig. 7.19, we can infer that the mixing process progressed with different efficiency in the three case studies. In particular, it can be said that the magma mixing efficiency increases for natural structures from Salina to Vulcano to Lesvos.

The method used above to quantify the mixing patterns in lava flows requires digital images in which the two magmas can be still well recognized. This is essential in order not to lose too much information during image segmentation (Perugini et al. 2003). If the compositional variation in the images is more blurred due to the onset of larger degrees of chemical diffusion, alternative methods need to be used.

Figure 7.20 shows some representative samples of mixing patterns from the pyroclastic deposit of the Pollara eruption (Island of Salina). In this case, the contacts between magmas are more fuzzy and it is difficult to establish objective criteria for image segmentation.

A possible technique that can be used to extract information from the mixing patterns of Fig. 7.20 is the power-spectrum method that we have discussed in Chap. 2. The method is applied here on compositional series extracted from digital images. As we have seen before, there is a clear correlation between the grey value of the images and their SiO_2 content. Therefore, compositional series represent the variation of SiO_2 across the samples.

A representative compositional series extracted from a natural sample is reported in Fig. 7.21a. We can see that concentration varies erratically defining up and down patterns presumably reflecting the presence of a large number of filaments of the two magmas. Figure 7.21b shows the corre-

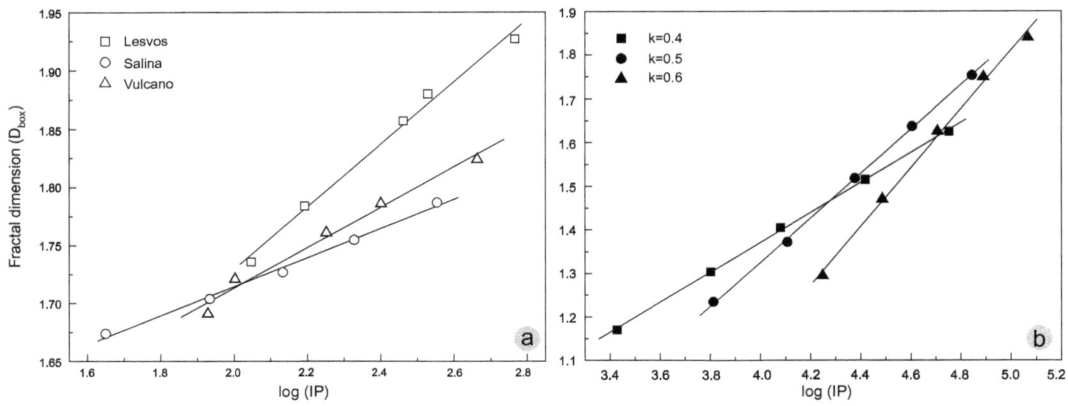

Fig. 7.19 Variation of the length of interfaces (IP) versus fractal dimension (D_{box}) for natural (**a**) and numerically reproduced flow structures (**b**)

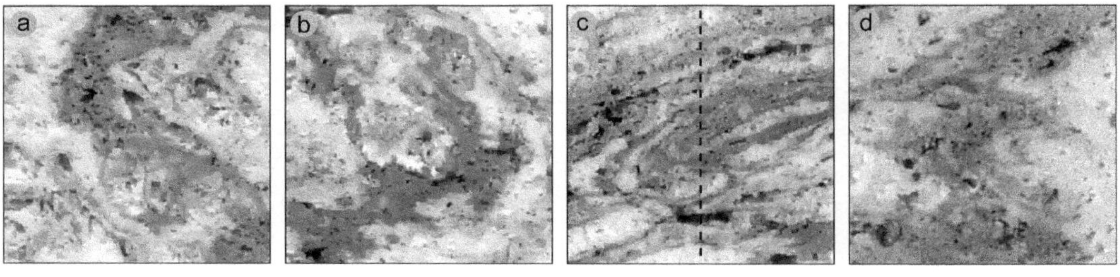

Fig. 7.20 Examples of magma mixing patterns from the Pollara pyroclastc deposit. In **c** is reported the transect used to extract one of the compositional time series analysed by the power spectrum method

sponding power-spectrum. Data follow a power-law distribution of concentration frequencies, as indicated by the fact that data are aligned along a linear trend (see Chap. 2). Note that, according to the discussion reported in Chap. 2, the plot shows that the compositional series is not random, but shows a clear scaling pattern supporting the fractal nature of the compositional field. The spectral slope is a measure of mixing intensity.

Spectral analysis was performed on ca. 200 samples and spectral slopes were estimated. Results are given in Fig. 7.22 where the frequency distribution of spectral slopes is reported. From the graph it is clear that spectral slopes vary among the different samples and are distributed in the range 1.50–1.80. In addition, they are normally distributed and the distribution is centered at ca. 1.66. This indicates that (i) the mixing process developed with different intensities in the natural samples and (ii) most of the mixing process developed with a mixing efficiency which is normally distributed throughout the magmatic system. It is interesting to compare the distribution of mixing efficiency recorded in natural samples (Fig. 7.21) and the distribution of the stretching field measured for the sine flow map (Fig. 4.2). The distribution of the stretching field is a measure of mixing efficiency in the whole system and the fact that it has approximately a normal distribution corroborates the idea that natural structures are the product of chaotic mixing processes.

7.3.3 Reproduction of Flow Structures Using High-Temperature Experiments

High-temperature experiments can be used to mimic the formation of flow structures in magmatic systems. As discussed in Chap. 5,

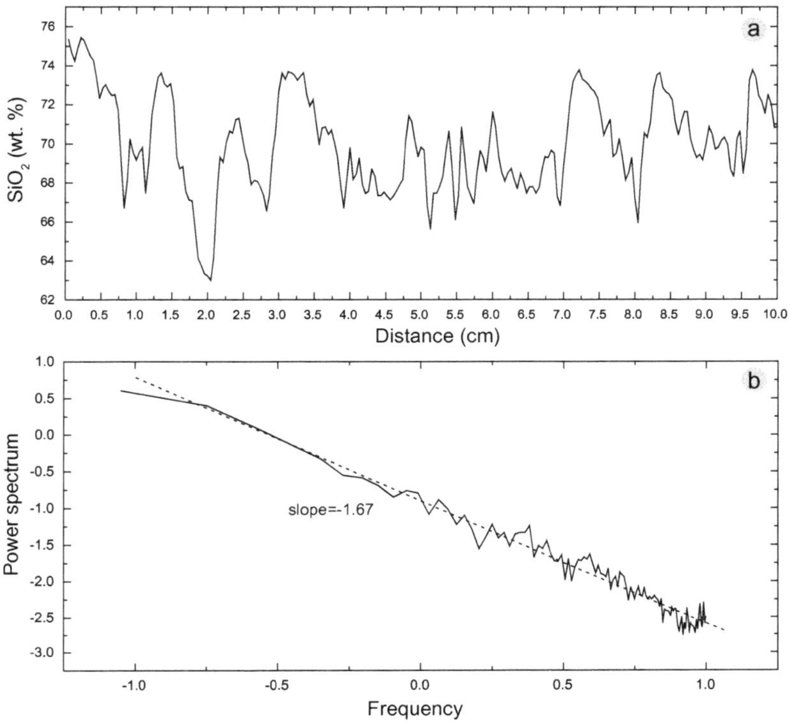

Fig. 7.21 **a** Compositional time series extracted along the transect shown in Fig. 7.20c; **b** power spectrum of the concentration variability. The presented spectrum results from fifty averaged compositional transects from the same image. The best-fit line (dashed) of data and the relative slope is given in the graph

an experimental device that can be used to this aim is the Chaotic Magma Mixing Apparatus (COMMA). Here we present some experimental results in order to shed further light on the formation of flow structures and their ability to modulate the compositional variability in the mixing system.

Figure 7.23a–c shows the outcome of a series of mixing experiments using as end-member magmas natural basaltic and rhyolitic melts from the Snake River Plain (USA; Morgavi et al. 2013; Morgavi et al. 2013). Three experiments were performed at different mixing times (Fig. 7.23), using the mixing protocol described in Chap. 5 (see Morgavi et al. 2013 for further details). The complexity of patterns generated by the application of the chaotic mixing protocol increases with increasing mixing time, as expected. Digital images of the experiments were converted into binary black and white picture (Fig. 7.23d–f), as for the natural mixing structures. The extent of mixing was measured estimating their fractal dimension (D_{box}).

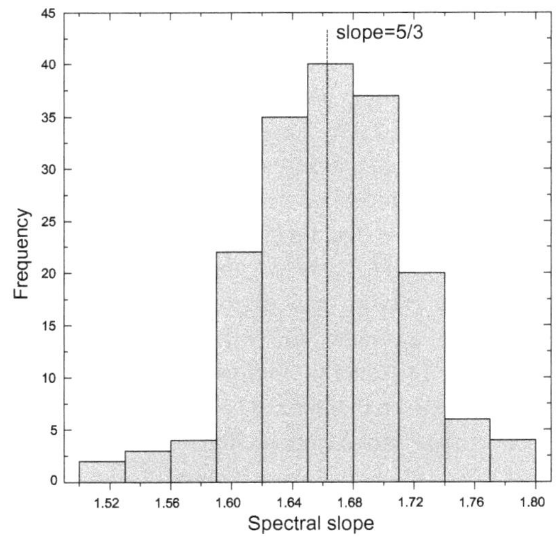

Fig. 7.22 Frequency histogram of spectral slopes estimated for the natural samples. The modal value corresponds to slopes of approximately −5/3

The plot of Fig. 7.24 shows the variation of D_{box} for the experimental mixing patterns as

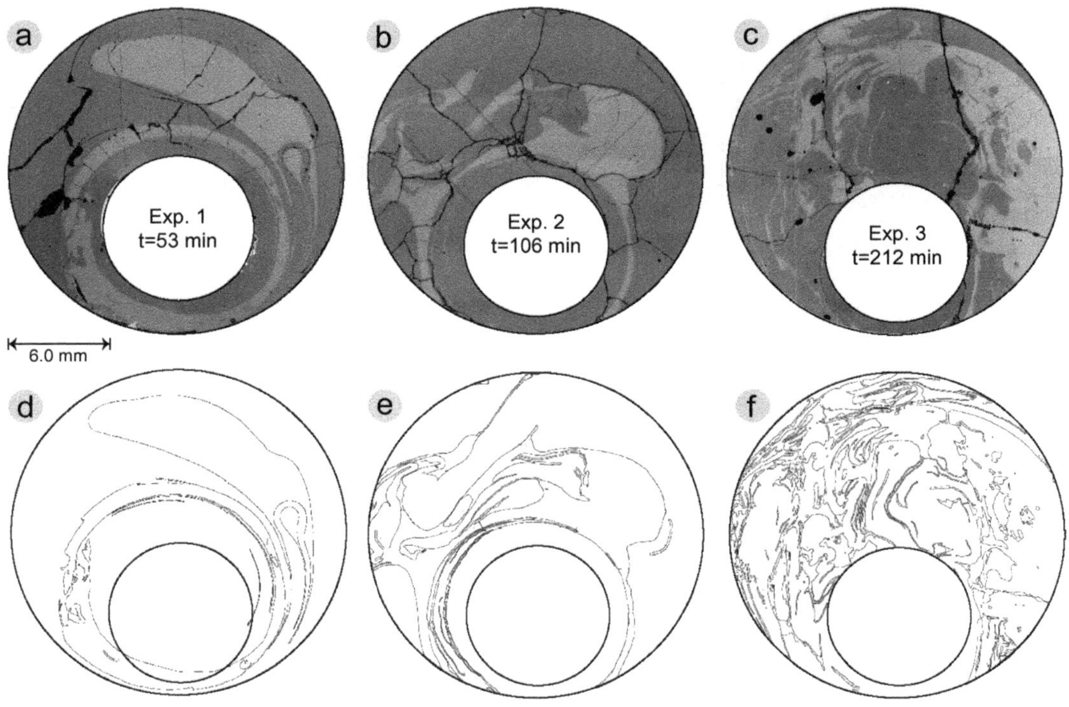

Fig. 7.23 **a–c** BSE images of three chaotic mixing experiments at different mixing times (t); **d–f** segmented images of the experiments showing the contact interfaces between the two melts. Modified from Morgavi et al. (2013)

a function of the relative length of interfaces [$\log(IP)$], as for natural and numerical flow structures (Fig. 7.19). These two variables appear linearly correlated, as already observed for natural samples and numerical simulations. Therefore, it can be said that the linear relationship between D_{box} and $\log(IP)$ is a "universal" character for magma mixing and might represent a good method to estimate the efficiency of mixing processes. In general, increasing mixing times correspond to an increase of complexity of mixing patterns that should be related to progressively more efficient chemical exchanges.

The compositional variation in the three experimental samples was extensively evaluated in Morgavi et al. (2013) and Morgavi et al. (2013). They analyzed compositional transect across the mixing structures reported in Fig. 7.23 and estimated the concentration variance for major elements with the passing of mixing time (see Chap. 4 for details).

In Fig. 7.25 the concentration variance (σ_n^2) of major elements is plotted as a function of the fractal dimension (D_{box}) of the mixing pattern at different times. The graph shows that σ_n^2 displays an exponential variation with the morphological complexity of the mixing pattern (i.e. the fractal dimension, D_{box}). An important result emerging from Fig. 7.25 is that the different chemical elements show different exponential relationships with D_{box}. This means that the different elements have different mobilities in the magmatic system, as already observed for both numerical simulations and high-temperature centrifuge experiments (Chaps. 4 and 5). A further, non-trivial result is that the extent of homogeneity of chemical elements correlates well with the morphology of the mixing pattern. As for the mobility of chemical elements, and for the end-members considered in the experiments, it increases in the following order: SiO_2, TiO_2, Al_2O_3, MgO, CaO, FeO_{tot}, K_2O and Na_2O.

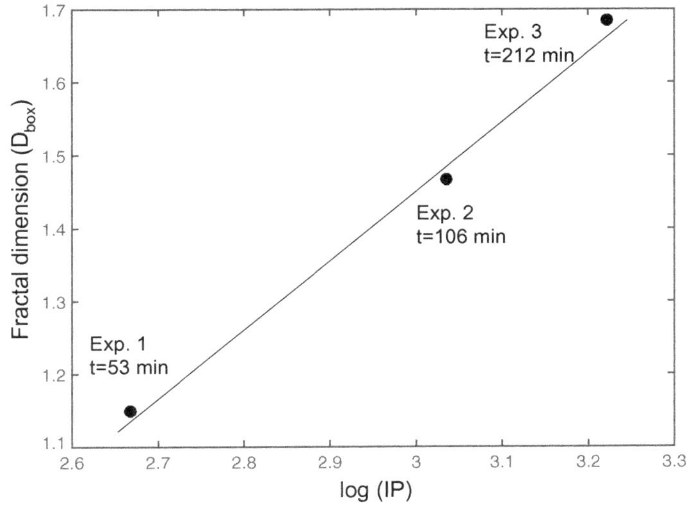

Fig. 7.24 Variation of the length of interfaces (IP) versus fractal dimension (D_{box}) for chaotic mixing experiments shown in Fig. 7.23

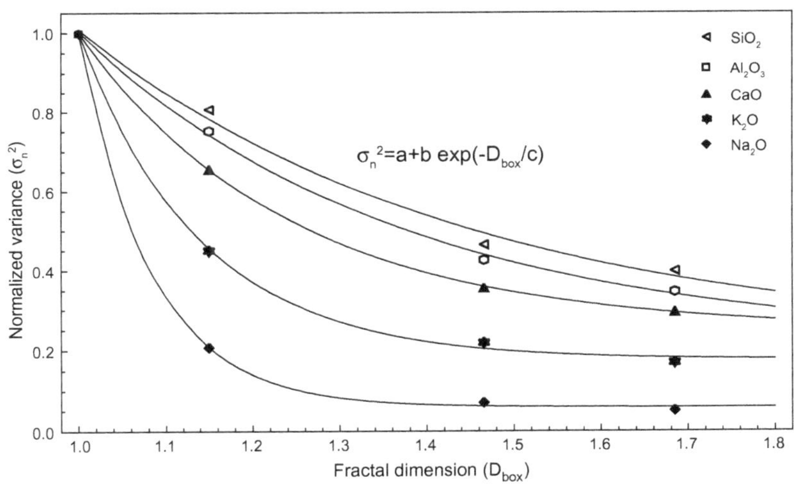

Fig. 7.25 Variation of concentration variance (σ_n^2) for major elements at different mixing times as a function of fractal dimension (D_{box}) for the mixing patterns in Fig. 7.23. The general exponential relationship used to fit data is also reported

7.3.4 Timing of Homogenization of Flow Structures

It is now worth discussing the possible timescales required for homogenization of flow structures, as we already did for magmatic enclaves. In the case of enclaves, we considered that they were trapped within regular regions since the beginning of the mixing process and remained so for the rest of the lifetime of the magmatic system. Regarding flow structures the situation tends to be more complex as they are the product of repeated stretching and folding dynamics that can potentially develop in the system from the beginning of mixing to the time at which the magmatic system is unable to flow (e.g. due to temperature drop, crystallization, increase of viscosity, etc.). Nevertheless, under some assumptions, we can attempt to estimate homogenization timescale also for flow structures.

Figure 7.26a shows a transect across a flow structure from the same lava flow in which the homogenization timescale was estimated for the magmatic enclave (Fig. 7.13). EMPA analyses were performed along the dashed line reported in the figure. Figure 7.26b reports the variation of Al_2O_3 along the transect. We can observe a correspondence between the mixing pattern and the compositional variation: the presence

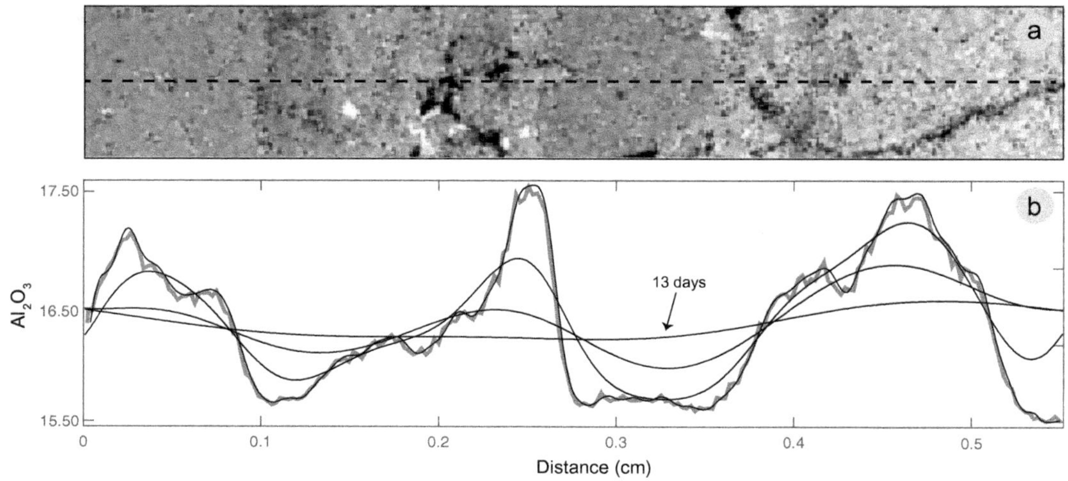

Fig. 7.26 **a** Back scattered electron images of flow structures constituted by alternating bands of trachyte and rhyolite from the Lesvos lava flow. The dashed line indicates the path followed by EMPA analysis; **b** Fourier series fitting of the variation of Al_2O_3 across the flow structure. The thick grey line represents the measured compositional variability. The variation of $C(x, t)$ against x at different times is also shown

of alternating portions of trachytic and rhyolitic glasses produces up and down fluctuations of Al_2O_3 concentration. As for the magmatic enclave (Fig. 7.13), the variation of Al_2O_3 concentration was fitted using the Fourier series technique whose polynomial expansion is used as the periodic function in the calculations of the diffusion patterns. The time required for complete homogenization was estimated using the same value of diffusion coefficient for Al_2O_3 used for the enclave. Here we are assuming that the system remains in static conditions after having reached the configuration shown in Fig. 7.26a and that no further stretching and folding dynamics develops in the system.

The time evolution of the diffusion profile is shown in Fig. 7.26b. As for the enclave, we consider that the system is homogeneous when the variation of concentration falls within the limits of measurements by EMPA. Results of the modeling indicate that the time required for the complete homogenization of the flow structure is of the order of 13 days. This timescale is ca. thirty times faster than the time required to homogenize the enclave (Fig. 7.13). This estimate is quite conservative because, as mentioned above, if the magmatic mass is not frozen by an eruption and continues to flow, a larger number of thinner filaments can be formed leading to homogenization timescales that can be even shorter.

From the results presented above, it is evident that magma mixing processes can produce, in the same system and at the same time, portions of magmas with very variable compositions. In particular, these processes can generate volumes of magmas which are completely homogenized (chaotic regions) coexisting with portions of magmas preserving their initial compositions (regular islands). These results are strictly related to the local structure of flow fields where chemical exchanges occur very differently depending on the extension of contact interfaces between magmas. These dynamical regions can proliferate throughout the magmatic mass from the metre to the micrometre length scale, because of the fractal nature of magma mixing patterns.

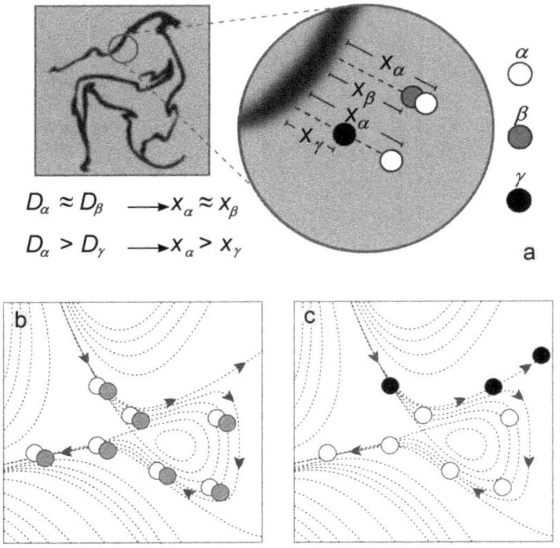

Fig. 7.27 Schematic illustration of the diffusive fractionation process. **a** portion of mixing system with flow structures of one magma (dark coloured) dispersed into another magma (light coloured). Elements with similar and different diffusion coefficient cover different distance (x) at the same time; **b–c** dispersion of diffusing elements superimposed on a hypothetical chaotic flow field. Elements having similar D remain one close to the other for some time during the mixing process while elements with different D strongly diverge

7.4 Diffusive Fractionation of Chemical Elements During Chaotic Mixing

7.4.1 Chemical Exchanges During Magma Mixing

The occurrence of volumes of melts generated by magma mixing having extremely variable compositions open several key questions. Assuming the magmatic system underwent chaotic mixing, can we still use the chemical composition of rock samples to infer the original composition of end-member magmas? Can these compositions be explained using classical geochemical models? These are not trivial questions because, for example, the geochemical fingerprint of igneous rocks is often taken as an indicator of the source rock and tightly related to the geodynamic setting.

We have seen in this and the previous chapters that the different chemical elements have different mobilities. This implies that at the same time the magmatic system is constituted by volumes of melts in which chemical elements are variably enriched (or depleted) depending on their ability to spread throughout the magmatic systems.

The occurrence of chaotic dynamics during mixing can also amplify this effect and can have a big impact on the modulation of the compositional fields in the magmatic system, triggering an additional petrological process known as "diffusive fractionation" (e.g. De Campos et al. 2011; Perugini et al. 2006; Perugini et al. 2008). As highlighted before, the contact area between interacting magmas in chaotic regions increases exponentially in time and chemical diffusion becomes more efficient as the length scale of filaments becomes progressively smaller. In addition, in the chaotic regions, nearby trajectories of the flow field diverge exponentially as mixing progresses in time (see Chap. 2).

We can visualize chemical diffusion as the process moving elements from a certain volume of magma to another, in agreement with the presence of a concentration gradient (Fig. 7.27). The distance (x) traveled by an element will depend on the diffusion coefficient (D): the larger D, the larger is the distance.

If two elements (say α and β) have a similar D, the distance they will travel in the same time span will be similar; on the contrary, if two elements (say α and γ) have different D the distance they will travel will be different (Fig. 7.27). This way melts with different compositions will be generated. Chaotic flow fields will distribute these volumes of melts. Since elements α and β have ca. the same D, their relative position will remain ca. the same for a certain time during advection (Fig. 7.27b). Contrarily, due to the larger difference in D, elements such as α and γ will suffer the effect of chaotic dynamics more rapidly (Fig. 7.27c). This process can be recognized as a variable correlation between pairs of elements in binary plots. Elements having similar D will be well correlated in inter-elemental plots defining ca. linear patterns, as it should be expected from a two end-member mixing process (Fig. 7.28a). On the contrary, for the same system, elements with

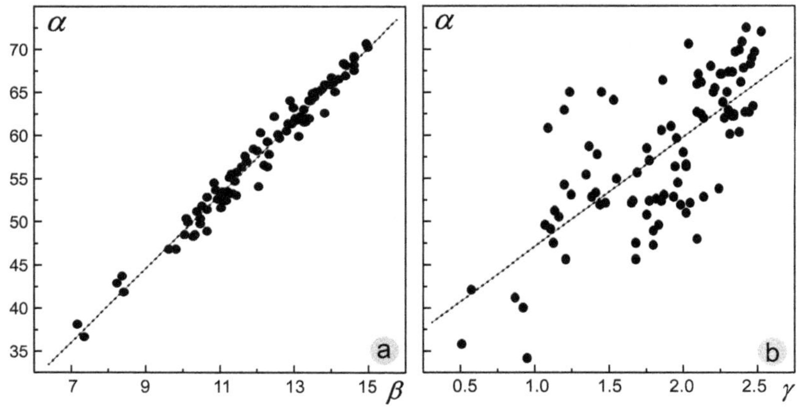

Fig. 7.28 Data from advection-diffusion numerical simulations (Chap. 4) showing the effect of diffusive fractionation on inter-elemental plots. Elements with similar diffusion coefficients will be linearly correlated, whereas, at the same time, elements with different diffusion coefficients will show complex patterns

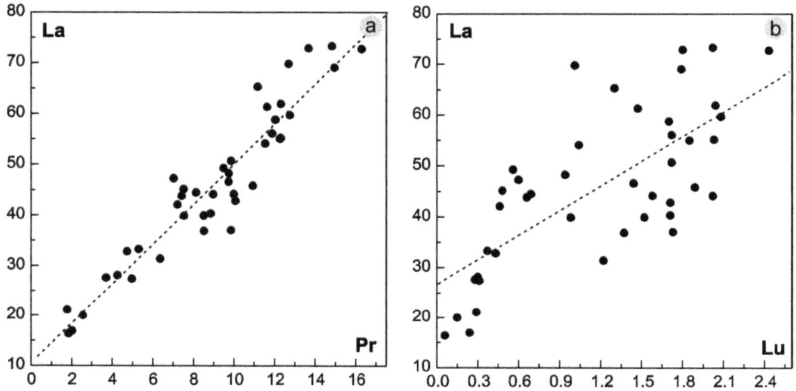

Fig. 7.29 Data from a lava flow erupted on the Island of Lesvos (Greece) displaying that elements with similar mobility are linearly correlated, whereas elements with different mobility are scattered

different D will show a progressive scattering as the difference in their D values increases (Fig. 7.28b).

Such a process can be recognized for natural samples (Perugini et al. 2006, Fig. 7.29), numerical models (Perugini et al. 2006; Fig. 7.28) and high-temperature magma mixing experiments (Morgavi et al. 2013; Fig. 7.30), suggesting that the evolution of chaotic mixing processes in space and time can produce a variety of compositional patterns in binary plots which, in many cases, cannot be modeled by the existing petrological models.

7.4.2 Rethinking Conventional Linear Mixing Models

One of the most striking results arising from results discussed above is that, during mixing, chemical elements experience a diffusive fractionation process. This process is considered the cause of deviations of chemical elements from linear variations in binary plots. Relying upon the assumption of linear trends during magma mixing (e.g. Fourcade and Allegre 1981) might generate aberrations leading to erroneous interpretations and misleading reconstruction of the end-members magmas. This, in turn, may have serious consequences on the identification of the source regions and their use to decipher the geological and geodynamical settings.

A quantitative evaluation of the failure of applicability of a linear two end-members mixing model can be done using the mixing test proposed by Fourcade and Allegre (1981). This is commonly used to decipher the possible occurrence of magma mixing processes (e.g. Janoušek et al. 1985; Solgadi et al. 2007). During this test the difference in all chemical elements between the felsic and the

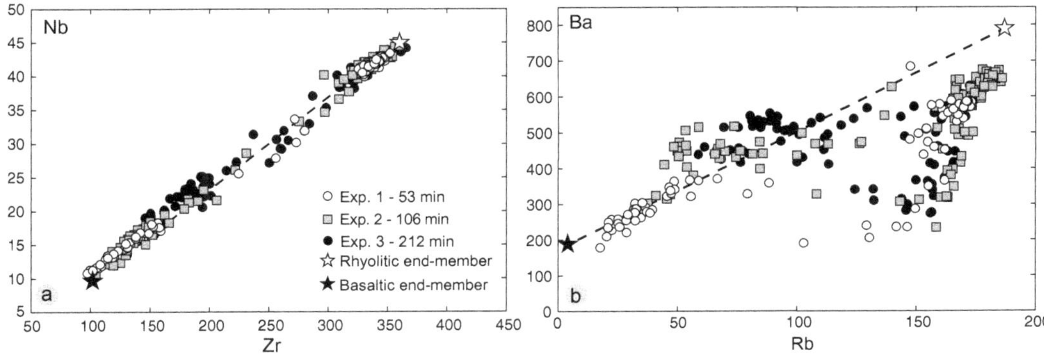

Fig. 7.30 Geochemical data from the mixing experiments shown in Fig. 7.23 showing that elements with similar and different mobility display linear and non-linear correlations in binary plots, respectively

mafic end-member (i.e. $C_F - C_M$) is plotted as a function of the difference in the same chemical elements in a supposed hybrid sample and the mafic end-member (i.e. $C_H - C_M$). Linear interpolation of data provides two different bits of information: (i) the correlation coefficient (r^2): this value indicates whether the hypothesized hybrid sample is actually the product of mixing (the larger the r^2 value, the better the evidence that the sample is a hybrid product of mixing); (ii) the slope of the linear fitting (x): this provides information on the fraction of the felsic end-member. The equation used in the mixing test derives from the linear mixing equation [$C_H = C_F x + C_M(1 - x)$]:

$$C_H - C_M = x\,(C_F - C_M) \qquad (7.5)$$

Linear interpolation of data in the $C_F - C_M$ versus $C_H - C_M$ space provides the values of r^2 and x. Before applying the test, the sample supposed to have the hybrid composition must be chosen. When the composition of the end-members and their relative fractions are known, the hybrid concentration for each element can be estimated with the two end-member mixing equation.

As an example, we can apply the mixing test using data from the experiments presented in Fig. 7.23 collected at different mixing times (Morgavi et al. 2013). From the available data set were selected those samples having the hybrid composition for a given chemical element. Figure 7.31a, c show the application of the mixing test considering the compositional variability of two major and trace elements for the long duration mixing experiment. Only a segment of the $C_F - C_M$ versus $C_H - C_M$ diagram can contain chemical elements for a two end-member mixing process; this is bounded by the lines with slope $x = 1.0$ (i.e. the system is constituted only by the rhyolitic end-member) and $x = 0.0$ (i.e. the system is constituted only by the basaltic end-member; Fig. 7.31a, c).

Considering major elements, it is shown that almost all elements fall in the graph domain which is compatible with a two end-member mixing process. For these elements the correlation coefficients (r^2) are very high. This result may indicate that the selected sample has reached the hybrid composition for all elements. However, more careful examinations show that the fraction of the felsic end-member (i.e. the slope x of the linear fitting) is, in some cases, inaccurate. This is shown in Fig. 7.31b where the fraction of rhyolitic end-member claimed by each element is reported. In particular, the value of x calculated for each element from the plot on the left side is reported as a "spider diagram" on the right side. The plot shows that, considering for example K_2O as the reference element, almost all the other elements show remarkable deviations. The same considerations can be made for the other major elements (see Morgavi et al. 2013 for further details).

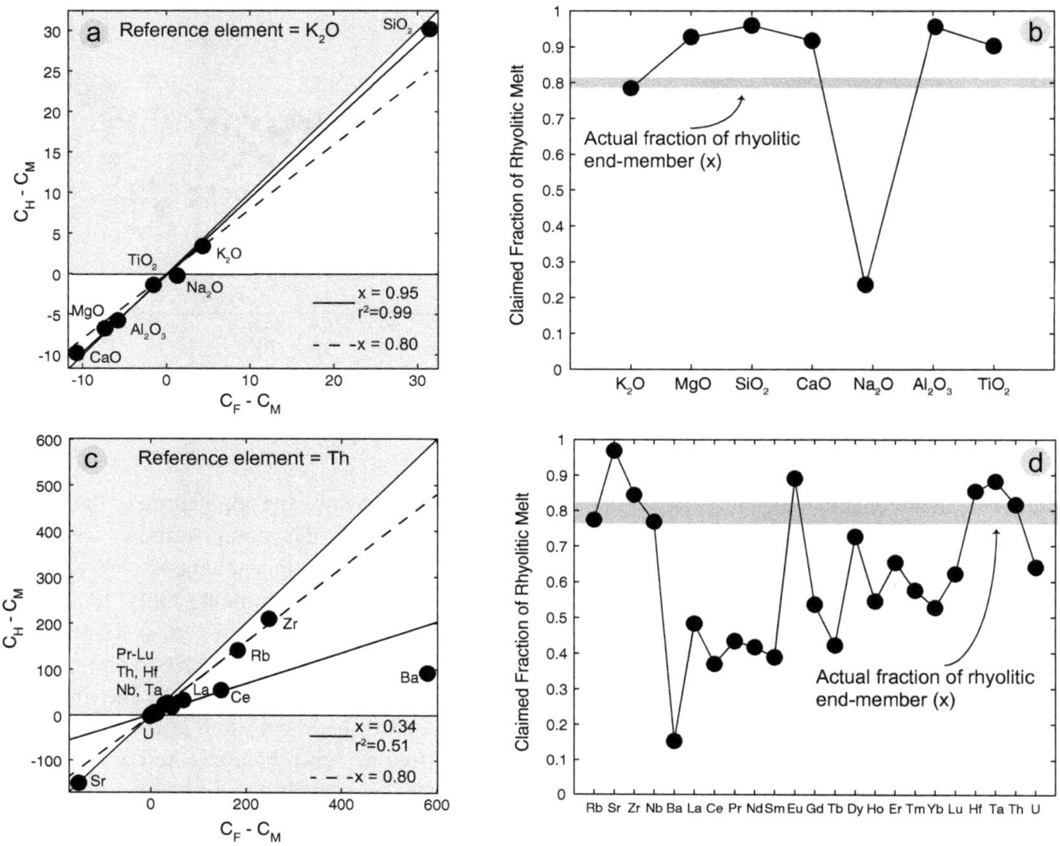

Fig. 7.31 **a** and **c** Application of the mixing test considering the hybrid concentrations of K_2O and Th; **b** and **d** graphs showing the fraction of felsic end-member indicated by each element (see text for details)

The plots for trace elements highlight even larger deviations, with most elements claiming for erroneous initial fractions of rhyolite. For example, considering Th as the reference element for the hybrid composition (Fig. 7.31c, d), all the other trace elements strongly deviate from the actual initial fraction of rhyolitic end-member. The other trace elements display a similar behaviour (Morgavi et al. 2013).

The failure in the application of the mixing test discussed above clearly indicates that a two end-member mixing conceptual model fails in explaining the compositional variability in magma mixing systems. This is the natural consequence of the fact that simple linear relations cannot model the time and space evolution of complex multi-component systems that, by their intrinsic nature, behave non-linearly due to the action of chaotic dynamics.

References

Albarede F (1995) Introduction to geochemical modeling. Cambridge University Press, Cambridge

Baker D R (1990) Chemical interdiffusion of dacite and rhyolite - anhydrous measurements at 1 Atm and 10 Kbar, application of transition-state theory, and diffusion in zoned Magma Chambers. Contrib Miner Pet 104(4):407–423

Baranger M, Latora V, Rapisarda A (2002) Time evolution of thermodinamic entropy for conservative and dissipative chaotic maps. Chaos Solitions Fractals 13:471–478

Christofides G et al (2007) Interplay between geochemistry and magma dynamics during magma interaction: an example from the Sithonia Plutonic Complex (NE Greece). Lithos 95(3–4):243–266. https://doi.org/10.1016/j.lithos.2006.07.015

Crank J (1975) The mathematics of diffusion. Clarendon Oxford

Davies GR et al (1994) Isotopic constraints on the production rates, crystallisation histories and residence times of pre-caldera silicic magmas, Long Valley, California. Earth Planet Sci Lett 125:17–37

De Campos CPP et al (2011) Enhancement of magma mixing efficiency by chaotic dynamics: an experimental study. Contrib Miner Pet 161(6):863–881. https://doi.org/10.1007/s00410-010-0569-0

De Rosa R, Donato P, Ventura G (2002) Fractal analysis of mingled/mixed magmas: an example from the Upper Pollara eruption (Salina Island, southern TyrrhenianSea, Italy). Lithos 65(3–4):299–311. issn: 00244937. https://doi.org/10.1016/S0024-4937(02)00197-4

Folch A, Marti J (1998) The generation of overpressure in felsic magma chambers by replenishment. Earth Planet Sci Lett 163(1–4):301–314. issn: 0012821X. 00196–4. https://doi.org/10.1016/S0012-821X(98)

Fourcade S, Allegre CJ (1981) Trace element behaviour in granite genesis: a case study the calc-alkaline plutonic association from the Querigut Complex (Pyrenees, France). Contrib Miner Pet 76:177–195

Furlong KP, Myers JD (1985) Thermal-mechanical modeling of the role of thermal stresses and stoping in magma contamination. J Volcanol Geotherm Res 24(1–2):179–191. issn: 03770273. 0273(85)90032–0. https://doi.org/10.1016/0377-

Grasset O, Albarede F (1994) Hybridization of mingling magmas with different densities. Earth Planet Sci Lett 121(3–4):327–332

Heath E et al (1998) Long magma residence times at an island arc volcano (Soufriere, St. Vincent) in the Lesser Antilles: evidence from 238U–230Th isochron dating. Earth Planet Sci Lett 160:49–63

Janoušek V et al (2000) Modelling diverse processes in the petrogenesis of a composite batholith: the central Bohemian Pluton, Central European Hercynides. J Petrol 41(4):511–543. issn: 1460–2415. https://doi.org/10.1093/petrology/41.4.511

Kuo CS et al (1997) Kinetics of spatially confined precipitation and periodic pattern formation. Physica A 239(1–3):390–403

Mandelbrot BB (1982) The fractal geometry of nature. W.H. Freeman

Morgavi D et al (2013) Interactions between rhyolitic and basaltic melts unraveled by chaotic mixing experiments. Chem Geol 346:199–212. issn: 00092541. https://doi.org/10.1016/j.chemgeo.2012.10.003

Morgavi D et al (2013) Morphochemistry of patterns produced by mixing of rhyolitic and basaltic melts. J Volcanol Geotherm Res 253:87–96. issn: 03770273. https://doi.org/10.1016/j.jvolgeores.2012.12.007

Morgavi D et al (2013) Time evolution of chemical exchanges during mixing of rhyolitic and basaltic melts. Contrib Miner Pet 166(2):615–638. issn: 0010–7999. https://doi.org/10.1007/s00410-013-0894-1

Perugini D, Petrelli M, Poli G (2008) A virtual voyage through 3D structures generated by chaotic mixing of magmas and numerical simulations: a new approach for understanding spatial and temporal complexity of magma dynamics. Vis Geosci 13(1):1–24. issn: 16102924 (ISSN). https://doi.org/10.1007/s10069-006-0004-x

Perugini D, Poli G (2004) Determination of the degree of compositional disorder in magmatic enclaves using SEM X-ray element maps. Eur J Mineral 16(3):431–442. issn: 09351221. https://doi.org/10.1127/0935-1221/2004/0016-0431

Perugini D, Poli G (2005) Viscous fingering during replenishment of felsic magma chambers by continuous inputs of mafic magmas: field evidence and fluid-mechanics experiments. Geology 33(1):5–8

Perugini D, Poli G, Mazzuoli R (2003) Chaotic advection, fractals and diffusion during mixing of magmas: evidence from lava flows. J Volcanol Geotherm Res 124(3–4):255–279. issn: 03770273. https://doi.org/10.1016/S0377-0273(03)00098-2

Perugini D et al (2003) Magma mixing in the Sithonia Plutonic Complex, Greece: evidence from mafic microgranular enclaves. Miner Petrol 78(3–4):173–200. issn: 09300708. https://doi.org/10.1007/s00710-002-0225-0

Perugini D, Petrelli M, Poli G (2006) Diffusive fractionation of trace elements by chaotic mixing of magmas. Earth Planet Sci Lett 243(3–4):669–680. issn: 0012821X. https://doi.org/10.1016/j.epsl.2006.01.026

Perugini D, Valentini L, Poli G (2007) Insights into magma chamber processes from the analysis of size distribution of enclaves in lava flows: a case study from Vulcano Island (Southern Italy). J Volcanol Geotherm Res 166(3–4):193–203. issn: 03770273. https://doi.org/10.1016/j.jvolgeores.2007.07.017

Perugini D et al (2008) Trace element mobility during magma mixing: preliminary experimental results. Chem Geol 256(3–4):146–157. issn: 00092541. https://doi.org/10.1016/j.chemgeo.2008.06.032

Petrelli M, Perugini D, Poli G (2006) Time-scales of hybridisation of magmatic enclaves in regular and chaotic flow fields: petrologic and volcanologic implications. Bull Volcanol 68(3):285–293. issn: 02588900. https://doi.org/10.1007/s00445-005-0007-8

Piochi M et al (2009) Constraining the recent plumbing system of Vulcano (Aeolian Arc, Italy) by textural, petrological, and fractal analysis: the 1739 A.D. Pietre Cotte lava flow. Geochem Geophys Geosyst 10(1). issn: 15252027. https://doi.org/10.1029/2008GC002176

Snyder D (2000) Thermal effects of the intrusion of basaltic magma into a more silicic magma chamber and implications for eruption triggering. Earth Planet Sci Lett 175(3–4):257–273

Snyder D, Tait S (1996) Magma mixing by convective entrainment. Nature 379(6565):529–531

Solgadi F et al (2007) The role of crustal anatexis and mantle-derived magmas in the genesis of synorogenic Hercynian granites of the Livradois area, French massif central. Can Mineral 45(3):581–606. issn: 00084476. https://doi.org/10.2113/gscanmin.45.3.581

Volpe AM, Hammond PE (1991) 238U-230Th-226Ra disequilibria in young Mount St. Helens rocks: time constraint for magma formation and crystallization. Earth Planet Sci Lett 107:475–486

8 The Fingerprint of Magma Mixing in Minerals

I know that in the study of material things number, order, and position are the threefold clue to exact knowledge.

D'Arcy W. Thompson

Abstract

Disequilibrium textures in minerals are common in igneous rocks and can be caused by variations in pressure, temperature, or composition that perturbed a pre-existing state of equilibrium of the magma body. Although the variation of intensive variables can be invoked as the triggering factor, there is evidence that disequilibrium did not propagate everywhere with the same intensity in the magmatic mass. This is supported by the evidence that crystals of the same phase, with extremely variable disequilibrium, occur at very short length scales, often even in the same thin section. In this chapter we present quantitative studies of mineral phases showing chemical-physical disequilibrium from different volcanic systems, both suggested to have experienced magma mixing. We will attempt to provide an explanation of the puzzling evidence of crystals showing extremely varied zoning patterns coexisting at very short length scale.

8.1 Introduction

Disequilibrium textures in minerals are common in igneous rocks (e.g. Hibbard 1994; Stamatelopoulou-Seymour et al. 1990; Wallace and Carmichael 1994; Umino and Horio 1998; Kuscu and Floyd 2001). They can be caused by variations in pressure, temperature, or composition that perturbed a pre-existing state of equilibrium of the magma body (e.g. Nixon 1988; Ortoleva 1990; Dobosi and Fodor 1992; Rutherford and Hill 1993; Slaby et al. 2011; Słaby et al. 2008). Although the variation of intensive variables can be invoked as the triggering factor, there is evidence that disequilibrium did not propagate everywhere with the same intensity in the magma body. This is supported by the evidence that crystals of the same phase, with extremely variable disequilibrium, occur at very short length scales (often even in the same thin section). A paradigmatic example was provided, for example, in Anderson (1984) who documented the simultaneous presence of plagioclases with vary variable oscillatory zoning.

D. Perugini, *The Mixing of Magmas*, Advances in Volcanology,
https://doi.org/10.1007/978-3-030-81811-1_8

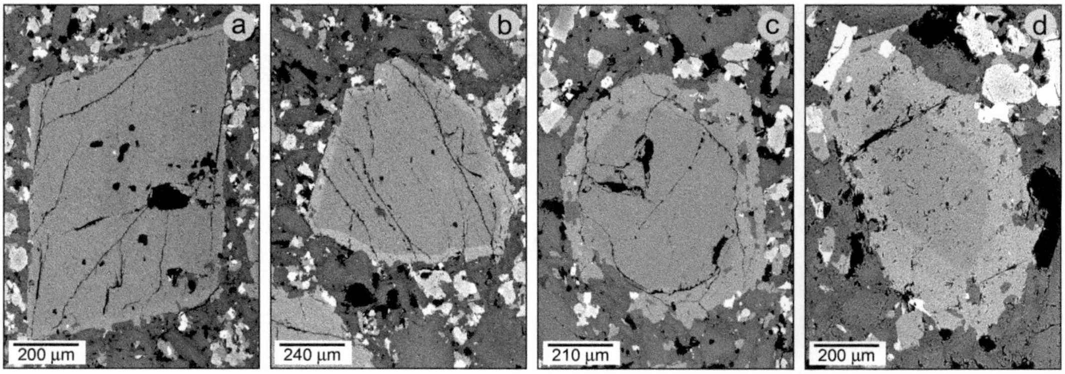

Fig. 8.1 Examples of pyroxene crystals (BSE images) from the Santa Venera lava flow (Mt. Etna, Italy) showing increasing thickness of growth rim and resorption cores

In this chapter we present quantitative studies of mineral phases showing chemical-physical disequilibrium from two different volcanic systems, both suggested to have experienced magma mixing. We will attempt to provide an explanation of the puzzling evidence of crystals showing extremely varied zoning patterns coexisting at very short length scale in the magmatic mass.

8.2 Compositional Zoning in Clinopyroxene Crystals

In this section we consider pyroxene crystals from the alkali basaltic lava flow of Santa Venera (Mt. Etna volcano, Italy) (e.g. Perugini et al. 2003).

Given that pyroxene can preserve disequilibrium textures for a long time (e.g. Thompson 1974; Shimizu 1990), we can regard it as an appropriate marker to study the development of disequilibrium in space and time in the magma body.

Pyroxene textures vary from euhedral and unzoned Cr-Al Di crystals to anhedral resorbed cores of Cr-Al Di with a rim of variable thickness of Al-Fe^{+3} Di composition (Fig. 8.1). This testifies for variable disequilibrium conditions suffered by Cr-Al Di during crystallization, causing resorption and allowing Al-Fe^{+3} Di to crystallize as rims. This crystal variability can be observed over a short length scale, of the order of a few centimetres.

Two representative crystals with different thickness of the Al-Fe^{+3} Di rim are shown in Fig. 8.2. Core-rim concentration variation of SiO_2 across these two crystals is displayed in Fig. 8.3. In general, pyroxenes can be classified into two main groups: (i) a first group showing core-rim irregular variations (Fig. 8.3a) and (ii) a second group displaying smooth and monotonous zoning patterns (Fig. 8.3b). Smooth compositional patterns characterize crystals with the tiniest Al-Fe^{+3} Di rims. On the contrary, irregular variations typically characterize pyroxenes having larger rims.

The thickness (area) of crystal rims was measured by image analysis (Fig. 8.4); Fig. 8.5a shows the distribution of rim areas in the form of a histogram. The plot displays three main crystal populations. The first one (P_1) identifies crystals with no rim; populations P_2 and P_3 are associated with crystals with progressively thicker rims.

Figure 8.5b reports the perimeter of resorbed pyroxene cores as a function of the rim area. The plot shows a non-linear relation between these two variables suggesting that crystallization occurred in an environment where disequilibrium conditions were highly fluctuating.

Perugini et al. (2003) suggested that magma mixing might explain the textural variability observed in Santa Venera rocks. As we have seen in the previous chapters, magma mixing can trigger the development of a remarkable heterogene-

Fig. 8.2 Pyroxene crystals (BSE images) across which core-rim compositional transects were measured (Fig. 8.3)

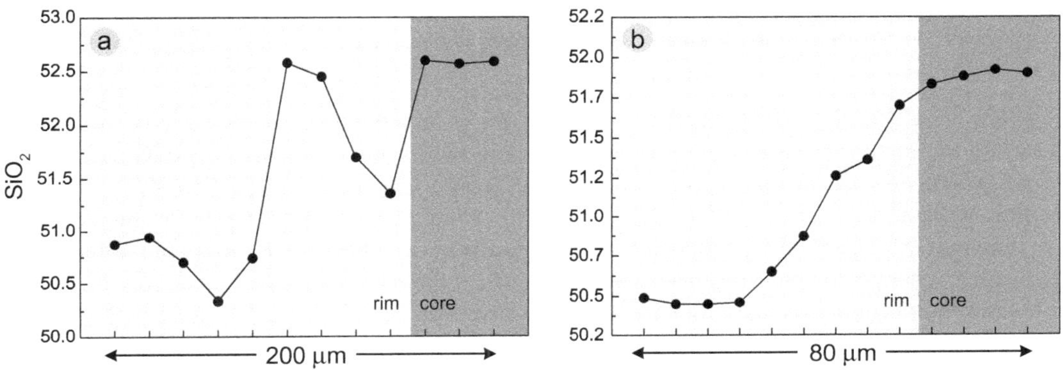

Fig. 8.3 SiO_2 core-rim compositional transects measured by EMPA on the crystals shown in Fig. 8.2

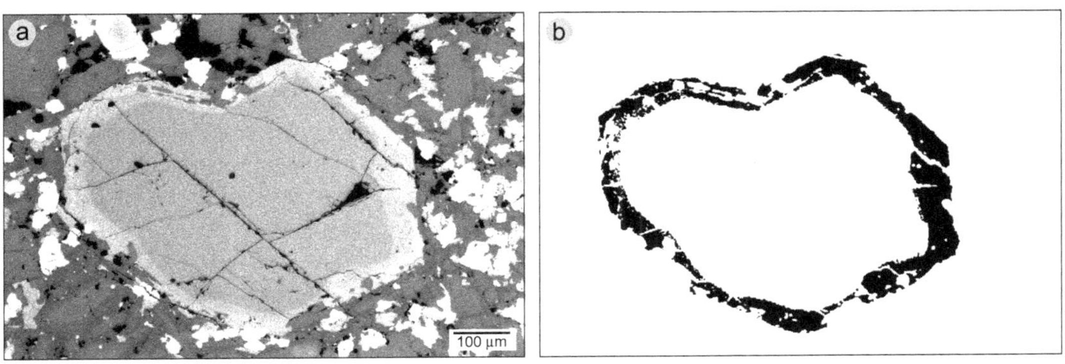

Fig. 8.4 Example of image segmentation of a pyroxene (a: BSE image) crystal for measuring the perimeter of the resorbed core and the area of the growth rim

ity propagating at many length scales in the magmatic mass due to the formation of fractal patterns. This can have an impact on the minerals that were already present in the end-member magmas and the newly formed ones (e.g. Hibbard 1994).

In the case of Santa Venera rocks, it was suggested that mixing occurred in a basaltic magma containing Cr-Al-Di crystals that was invaded by a more evolved magma (Perugini et al. 2003). Cr-Al Di suffered the disequilibrium triggered by

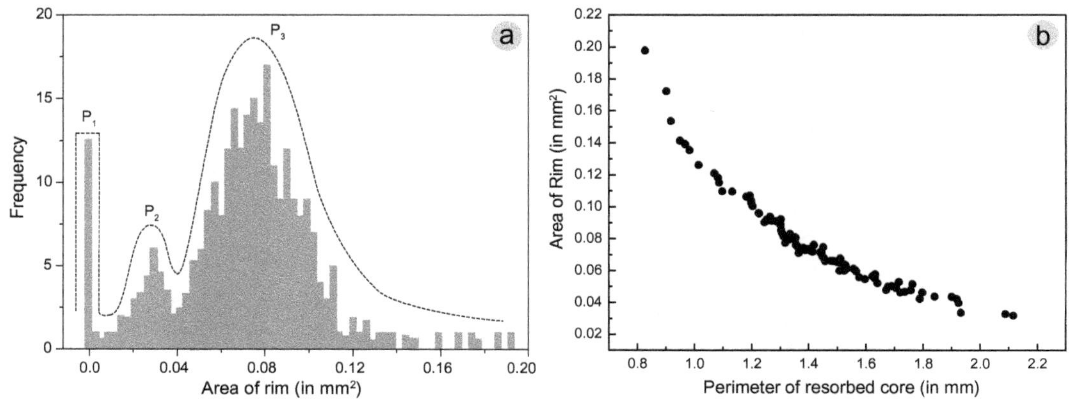

Fig. 8.5 **a** Frequency histogram of rim area for the pyroxenes from the Santa Venera lava flow (Mt. Etna, Italy); three crystal populations (P) are indicated. **b** Variation of the perimeter of the resorbed cores as a function of the rim areas

magma mixing. The thermodynamic variations induced by the mixing event caused resorption of Cr-Al Di. These acted as nuclei for the crystallization of the Al-Fe^{+3} Di.

Although in general this explanation seems reasonable, it does not explain the extremely variable resorption of pyroxene cores and the formation of three distinct populations of crystals with different rim thicknesses. Further, it does not explain the occurrence of extremely variable pyroxene crystals at very short length scale (Fig. 8.1).

We can attempt addressing these issues using numerical simulations of magma mixing in which minerals phases are allowed to grow. In doing this we use the advection-diffusion numerical scheme used in Chap. 4 (i.e. the sine flow map coupled with chemical diffusion).

At the beginning of the simulation, a large number of pyroxene nuclei are randomly distributed in the system domain (Fig. 8.6a). The growth of each nucleus is a function of the concentration of the nutrient element in its neighbourhood. The growth rate ψ_n of pyroxenes is calculated at each iteration of the numerical simulation as:

$$\psi_n = c_{(i-1)(j+1)} + c_{i(j+1)} + c_{(i+1)(j+1)} + c_{(i+1)j} + c_{(i+1)(j-1)} + c_{i(j-1)} + c_{(i-1)(j-1)} + c_{(i-1)j} \quad (8.1)$$

where c_{ij} is the value of each cell in the surrounding of each crystal. The total growth rate ψ_{tot} at the end of the simulation is given by the summation of all individual ψ_n and can be considered as a proxy for the rim area of crystals measured in natural rocks (Perugini et al. 2003).

Simulations were performed using a constant diffusion coefficient ($D = 0.4$) and different values of the control parameter k (k controls the relative size of regular and chaotic regions in the system; see Chap. 4) of the sine flow map. Here we present the results of numerical simulations performed using percentages of the incoming magma equal to 25 %. More comprehensive discussion considering different percentages is given in Perugini et al. (2003).

Frequency histograms of ψ_{tot} are presented in Fig. 8.7 for some representative simulations. In detail, Fig. 8.7a–c report the outcomes of simulations with values of the control parameter k equal to 0.4, 0.5 and 0.6, respectively.

For low values of k ($k = 0.4$ and $k = 0.5$) the histogram is constituted by three crystal populations (P_1, P_2 and P_3). As k increases ($k = 0.6$), the populations P_1 and P_2 disappear. Note that increasing k from 0.4 to 0.5 (Fig. 8.7a, b) populations P_1 and P_2 decrease, while population P_3 increases.

Being k a parameter controlling the relative amount of chaotic and regular regions in the mixing system, results suggest that populations P_1 and P_2 can be interpreted as due to pyroxene growth inside regular islands. Consequently, it can be inferred that the population P_3 is caused by crystal

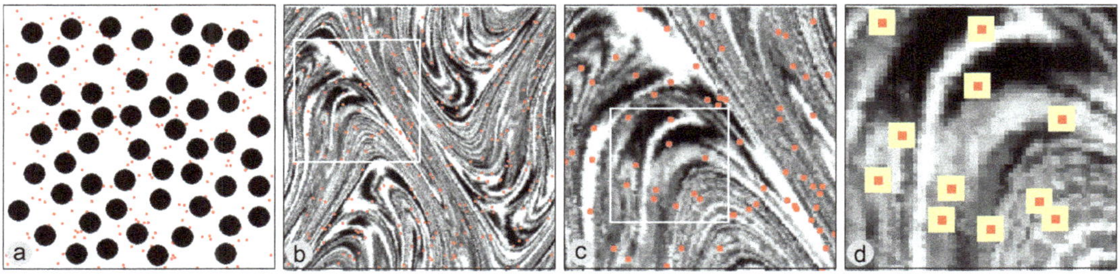

Fig. 8.6 Numerical simulation of the growth of pyroxenes in the magma mixing system; **a** random distribution of crystal nuclei inside the system; **b** snapshot of the numerical system at the sixth iteration of the mixing process; **c** and **d** magnifications of the system displaying the numerical procedure to calculate the growth potential of pyroxenes. In **d** the neighborhood of each cell in which concentration values are monitored is shown

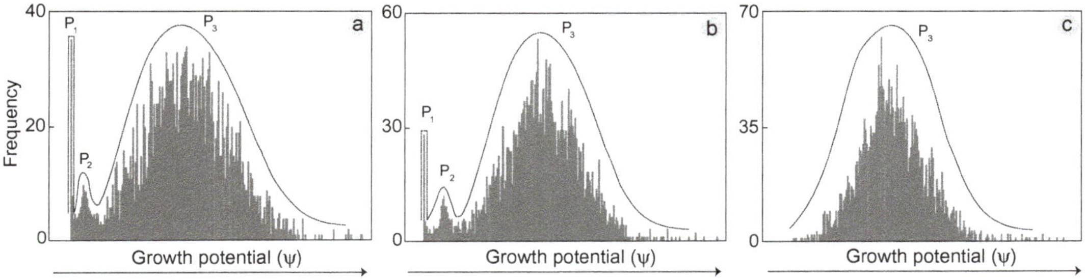

Fig. 8.7 Frequency histograms of ψ_{tot} for some representative simulations with values of the control parameter k equal to 0.4, 0.5 and 0.6

growth within the chaotic region. We have seen in Chaps. 4 and 6 that the transfer of concentration is very slow in regular islands, whereas chemical exchanges are much more efficient in chaotic regions.

The comparison of the histogram of natural data (Fig. 8.5a) with those resulting from the numerical simulations in which both regular islands and chaotic regions coexist (Fig. 8.7a, b) shows a striking similarity. In both the natural and simulated crystals three populations are present indicating that the numerical simulations can capture the general features of the studied outcrop. Crystal population P_1 and P_2 in natural rocks can be interpreted as related to those crystals the were trapped inside regular regions and that suffered none or little disequilibrium; crystal population P_3 can be interpreted as caused by disequilibrium of pyroxenes in the chaotic region.

Numerical simulations also offer the opportunity to explore the core-rim compositional variability for simulated crystals.

Figure 8.8 displays the compositional variation for two representative crystals that have grown into the regular and the chaotic region. The crystal in the regular region shows a smooth core-rim compositional variation. This is tightly related to the structure of the flow fields. In particular, regular islands behave as barriers to the transfer of chemical elements. In these conditions, pyroxenes deplete monotonously nutrient elements inducing the formation of smooth core-rim compositional patterns. The crystal in the chaotic region shows a more irregular compositional pattern. This is due to the fact that in chaotic regions the transfer of nutrients occurs more erratically causing the growth of pyroxenes to be less predictable and producing irregular core-rim compositional patterns.

The core-rim compositional patterns obtained from the numerical simulations and those observed in natural samples highlight a strong similarity, further supporting the use of the numerical simulation in catching the essential aspect of the natural process.

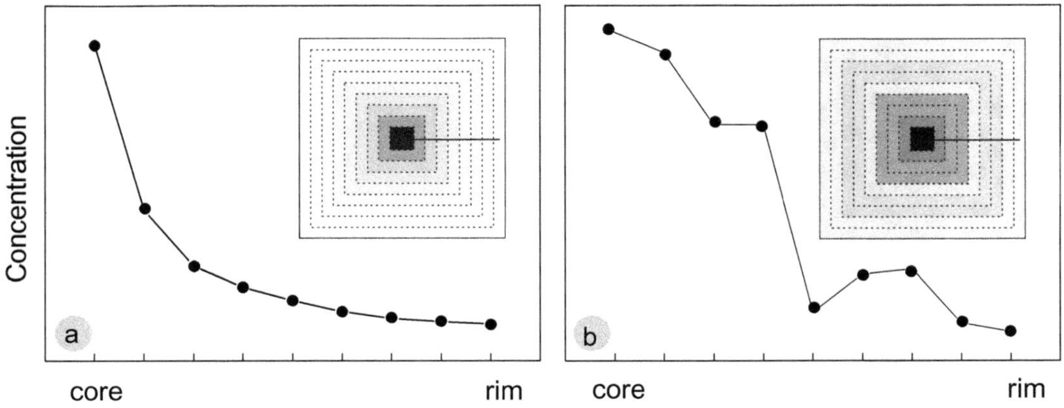

Fig. 8.8 Core-rim compositional transects for pyroxene crystals in **a** regular region and **b** chaotic region, resulting from the numerical simulation. Inset in each plot shows the compositional zoning in term of growth potential (different gray shades)

Resuming, we can hypothesize that pyroxene crystals inside the regular regions are forced to remain in restricted regions of the magmatic system (Fig. 8.9b). Here, they experience minimal effects of the thermodynamic disequilibrium induced by magma mixing (lower resorption, lower growth and smoother core-rim compositional variations). Contrarily, crystals in the chaotic regions are transported more erratically by flow fields (Fig. 8.9c), possibly encountering larger thermodynamic disequilibrium conditions. Here, changes in the scalar intensive variables (including concentration) are much more efficient, abrupt and unpredictable, leading to larger resorption, larger growth and more irregular core-rim compositional variations.

8.3 Oscillatory Zoning in Plagioclase Crystals

Oscillatory zoning in plagioclase crystals represents a magnificent archive of information (e.g. Anderson and Anderson 1984; Pearce and Kolisnik 1990; Ginibre et al. 2002; Wallace and Bergantz 2002). This, combined with the fact that this mineral phase is virtually ubiquitous in igneous rocks, makes plagioclase a formidable recorder of petrological processes to derive information on both the volcanological and petrological evolution of magmatic systems (e.g. Blundy and Shimizu 1991; Davidson and Tepley 1997; Stewart and Fowler 2001; Couch et al. 2001).

Here, we focus on the analysis of oscillatory zoning in plagioclases from three lava flows with HK-calc-alkaline affinity belonging to Monte Campanile and Porto (CP), San Rocco and Piano (SR), and Laghetto (LG) volcanites erupted on the Island of Capraia (Italy). These rocks have been interpreted as the result of petrological processes in which magma mixing dominated (Perugini et al. 2005; Perugini et al. 2006).

Figure 8.10 shows representative crystals from the lava flows. The zoning is characterized by different An/Ab ratio. In the BSE images of Fig. 8.10, zones enriched in An appear in the light grey colour; dark grey zones, instead, contain larger abundances of the Ab component.

Ginibre et al. (2002) showed that the grey shades of the individual zones in BSE images correlates linearly with their An content. Figure 8.11 shows that this relation also holds for the plagioclase crystals considered here. This being so, compositional series can be extracted from the oscillatory zoning of crystals (Fig. 8.12). An example of this procedure for a representative crystal (Fig. 8.12a) is shown in Fig. 8.12b.

Representative compositional series from the three lava flows are shown in Fig. 8.13. The series show continuous fluctuations of the Mol.%

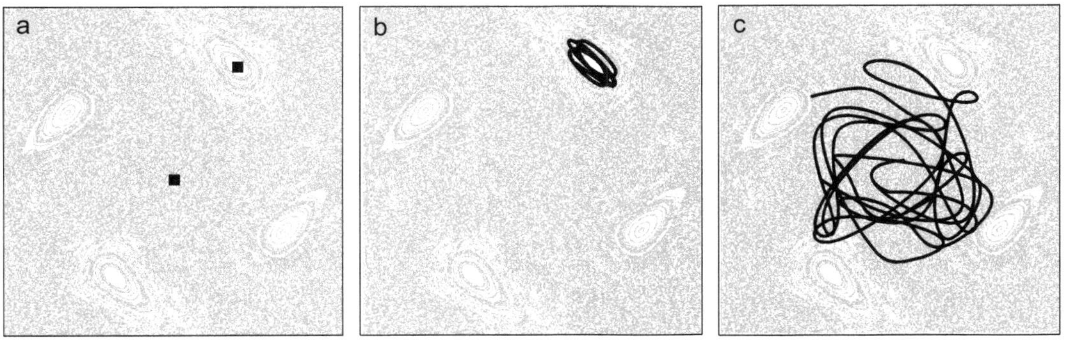

Fig. 8.9 **a** Flow field in the mixing system for $k = 0.4$ showing the position of two pyroxene crystals in the regular and the chaotic regions; **b** and **c** paths described by the two pyroxenes while transported in the regular and the chaotic region

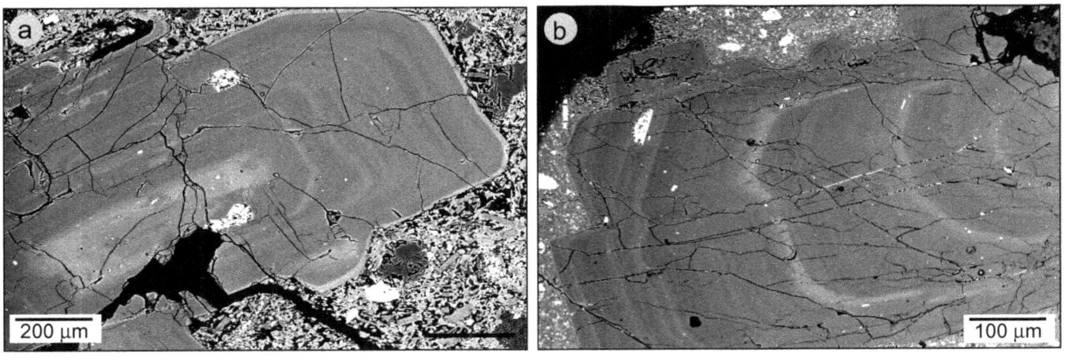

Fig. 8.10 Examples of plagioclase crystals (BSE images) from the Island of Capraia (Italy) showing the occurrence of oscillatory zoning patterns

An content, where short wavelength and high frequency variations are superimposed to long-term variations.

The quantitative analysis of compositional series was performed using a statistical technique known as Detrended Fluctuation Analysis (DFA; e.g. Kantelhardt et al. 2001; Hu et al. 2001). DFA can help in discovering correlations in sequences of data and, as such, it might be useful to derive information about the compositional fluctuations across plagioclase crystals. This, ultimately, will provide insights onto the processes that acted to generate them.

DFA involves different steps. Given a sequence of data (Fig. 8.14a; i.e. the compositional variation across the oscillatory zoning) $y(i), i = 1, \ldots, N$, where N is the length of the sequence, the sequence mean is calculated as

$$\bar{y} = \frac{1}{N}\sum_{j=1}^{n} y(i) \tag{8.2}$$

The integrated sequence $y(i), i = 1, \ldots, N$ (Fig. 8.14b), is then obtained as

$$x(i) = \sum_{j=1}^{i}[y(i) - \bar{y}], i = 1, \ldots, N \tag{8.3}$$

The integrated sequence $x(i)$ is divided into intervals of equal size n (Fig. 8.14c). A linear function $[x_{lin}(i, n))]$ interpolates the sequence in each interval. The linear interpolation defines the "local" trend in each interval.

The fluctuation sequence (Fig. 8.14) is, hence, calculated as

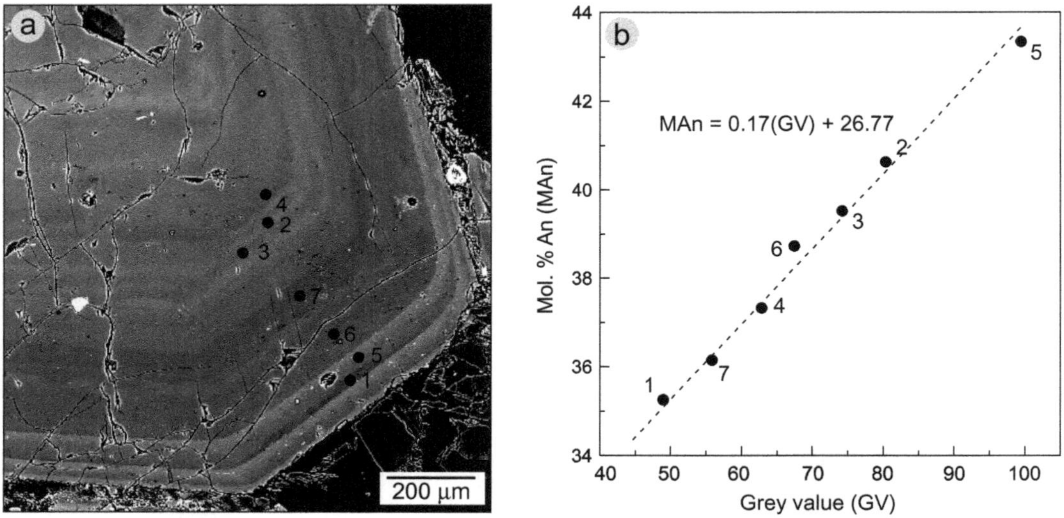

Fig. 8.11 **a** BSE image of plagioclase crystal across which EMPA analysis were performed (black dots indicate EMPA spots); **b** linear relationship between the grey value of the BSE image and the Mol.% An content

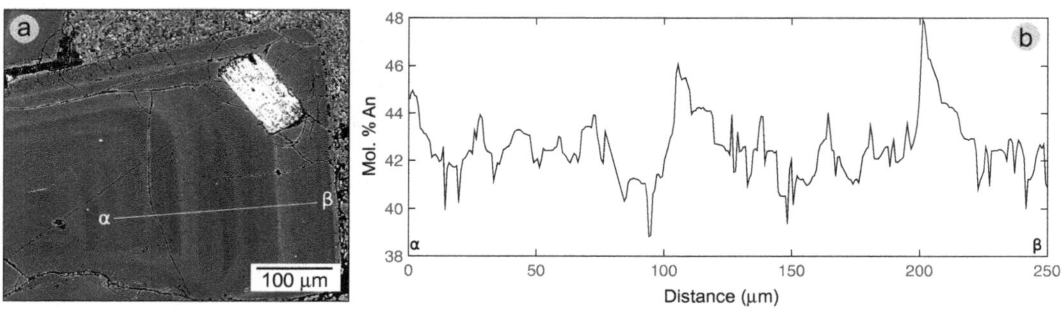

Fig. 8.12 **a** Representative plagioclase crystal (BSE image) across which a compositional time series is extracted (**b**)

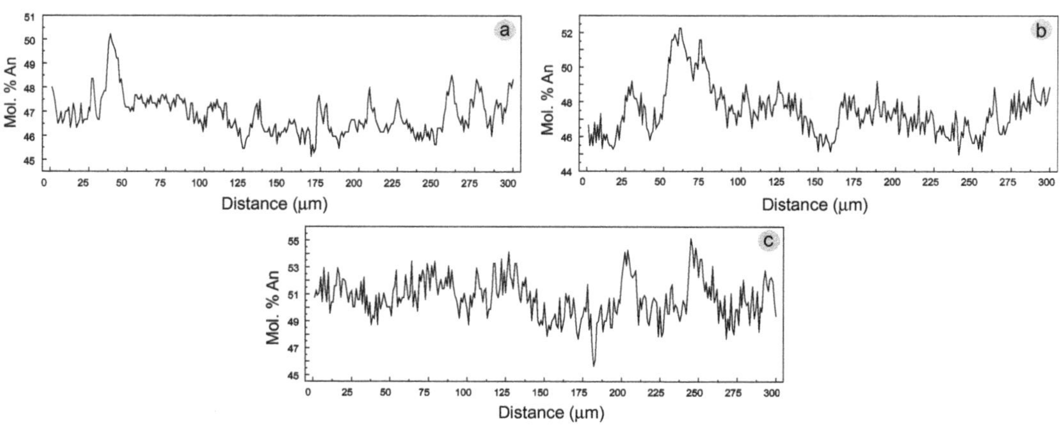

Fig. 8.13 Examples of compositional series extracted from three representative plagioclase crystals from SR (**a**), CP (**b**) and LG (**c**) lava flows

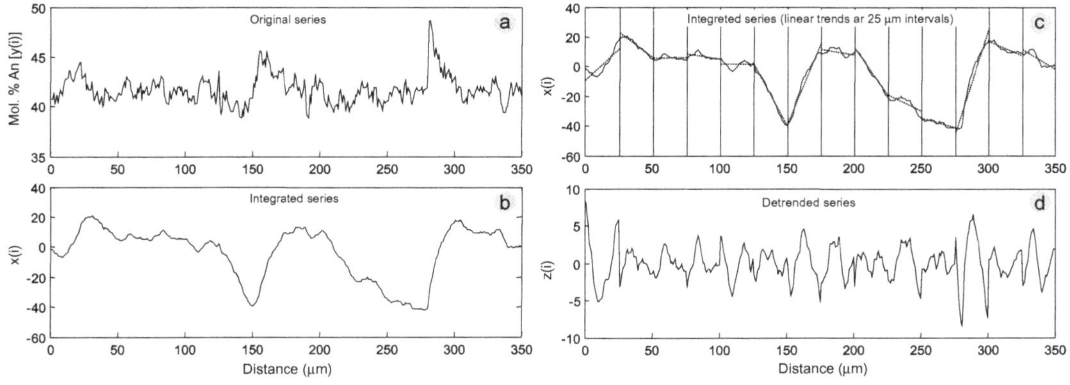

Fig. 8.14 Illustration of the Detrended Fluctuation Analysis (DFA) applied to a compositional series extracted form a natural plagioclase crystal

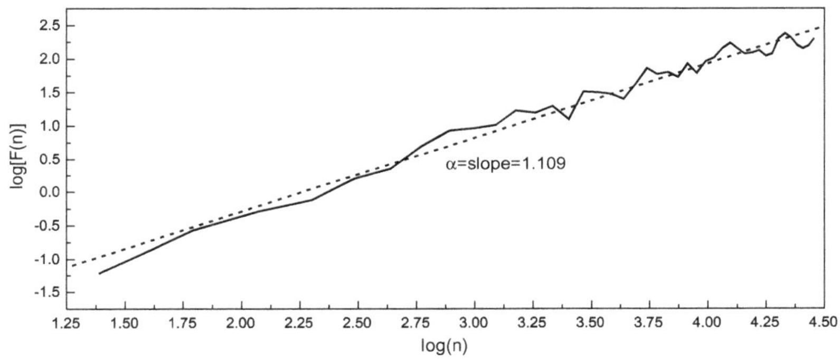

Fig. 8.15 Variation of $\log[F(n)]$ versus $\log(n)$ for the compositional series shown in Fig. 8.14a. The scaling exponent α is the slope of the linear fitting of data

$$z(i,n) = x(i) - x_{lin}(i,n), i = 1, \ldots, N \quad (8.4)$$

The root-mean squared value of the sequence $z(i,n)$ (Fig. 8.14d) is the fluctuation function $F(n)$:

$$F(n) = \sqrt{\frac{1}{N}\sum_{j=1}^{N} z(i,n)^2} \quad (8.5)$$

The procedure is repeated for a large range of intervals n. If the sequence defines a power-law the following relation is satisfied:

$$F(n) = n^{\alpha} \quad (8.6)$$

where α is the scaling exponent, which is estimated by linear interpolation of the $\log[F(n)]$ versus $\log(n)$ plot (Fig. 8.15).

α is related to the autocorrelation of the sequence. $\alpha = 0.5$ means no correlation and the sequence can be regarded as having properties of standard white noise (see Chap. 2). If $\alpha < 0.5$ the sequence is anti-persistent (i.e. an increment in the sequence is very likely to be followed by a decrement, and vice versa). If $\alpha > 0.5$ the sequence is persistent (i.e. an increment is very likely to be followed by an increment, and vice versa). Other values of α correspond to specific processes. For example, $\alpha = 1.5$ indicates Brownian motion; $\alpha = 1.0$ suggest $1/f$ noise. Further details on the meaning of α can be found in (Kantelhardt et al. 2001; Hu et al. 2001; Peng et al. 1994). Being α a measure of the autocorrelation of the data sequence, it can be used in the attempt to decipher the petrological message stored in plagioclase zonings and the processes responsible for their formation.

DFA was applied to plagioclase compositional series from the three lava flows and the value of the scaling exponent α was estimated. The results

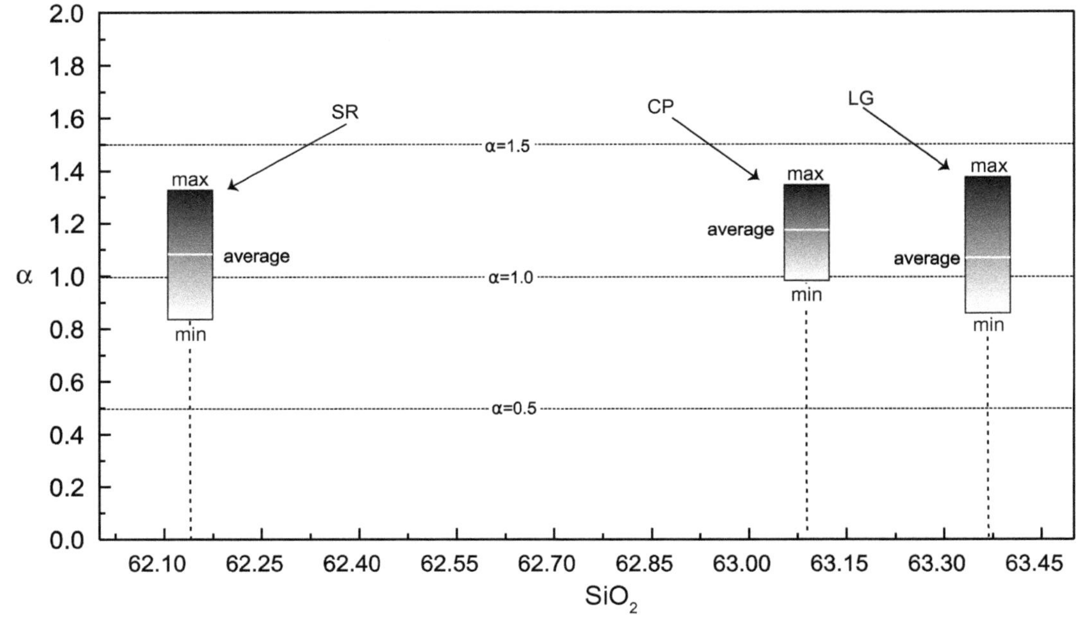

Fig. 8.16 Variability of α values for the plagioclase oscillatory zonings from the three lava flows plotted as a function of their average SiO_2 content. For each lava flow a box containing maximum (max), minimum (min) and average value (white line in the box) of α is reported. Lines for α equal to 1.5, 0.5 and 1.0 are also shown

are shown in the plot of Fig. 8.16 where α values are reported as a function of the average SiO_2 content of lava flows. α values display a quite narrow variability with average values clustering around one. Remarkably, no single oscillatory pattern is compatible with white noise ($\alpha = 0.5$) leading us to exclude that the zoning is the result of random processes. In addition, the fact that most α values are close to one indicates that oscillatory patterns can be interpreted as $1/f$-noise signals.

$1/f$-noise is typically taken as an evidence of Self-Organized Criticality (SOC; Bak et al. 1987; Bak et al. 1988) and many natural processes are shown to display this kind of behaviour (e.g. Bak and Tang 1989; Hergarten 2002).

Self-organized criticality is a phenomenon which is shown by systems reaching a critical state by their intrinsic dynamics. A paradigmatic example is a sand pile. As the pile grows, the pile itself become unstable and avalanches are triggered. Bak (1990) has shown that $1/f$-noise characterizes those systems whose dynamics evolve close to a threshold of instability (i.e. SOC). SOC, therefore, is a characteristic state due to self-organization of the system at the boundary between chaos and stability. Changes in the system are not smooth and continuous but occur by catastrophes. The fact that compositional series are constituted by segments of zoning with small Mol.% An variations followed by sharp increments and that their α values tend to cluster around $\alpha = 1.0$ might support the idea that they formed in magmatic system characterized by a sequence of critical states.

A simple dynamical system can be used to better understand how self-organized criticality may provide an explanation for plagioclase oscillatory zoning. It is known as the "logistic map" (e.g. May 1976; Turcotte 1992).

The formulation of the logistic map is:

$$X_{t+1} = rX_t\,(1 - X_t) \quad (0 < X_t < 1, 0 < r \leq 4) \tag{8.7}$$

where X_t and X_{t+1} are the states of the system at time t and $t + 1$, respectively. r is the control parameter of the map.

The map shows a rich spectrum of behaviours for different values of the control parameter r. This

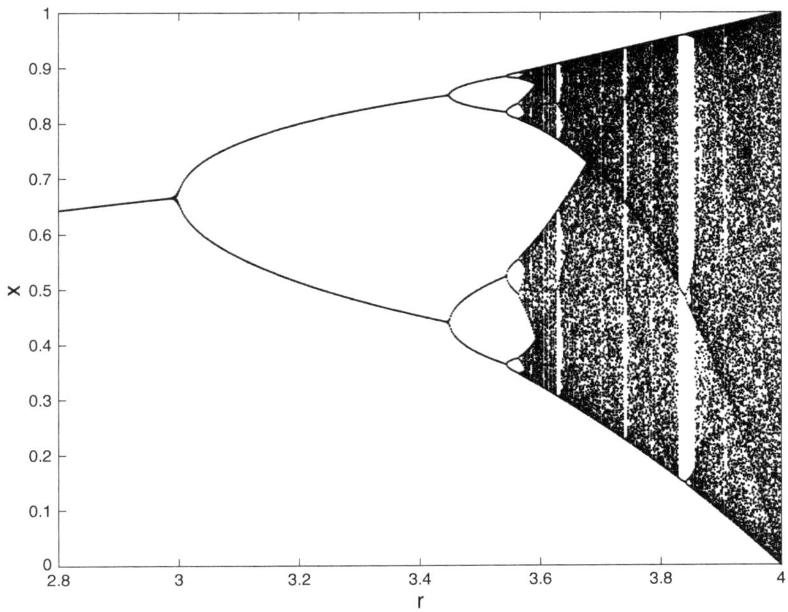

Fig. 8.17 Bifurcation diagram of the logistic map showing the general behavior of the dynamical system for different values of the control parameter r

can be visualized in the bifurcation diagram shown in Fig. 8.17, where the x-axis reports the values of r and the vertical axis shows the possible values for X_t.

Broadly speaking, the map is constituted by two main dynamical regions. The first is characterized by continuous lines and corresponds to stable dynamics (Fig. 8.17). The second is represented by dotted areas, corresponding to chaotic dynamics (Fig. 8.17). In some way this is an example in $1D$ of coexistence, in the same dynamical system, of chaos and regularity, a feature which we have been extensively discussed.

The two dynamical regions are barely separable. For example, regions of stability can be seen in the chaotic region on the right side of the plot close to the value of $r = 3.83$. There are an infinite number of stable regions intermingled with the chaotic regions.

As our interest is to study the system at a critical threshold, we use a value of $r = 3.9601$ laying at the edge of a chaotic region. The map was iterated and Fig. 8.18a shows the resulting data sequence.

Detrending fluctuation analysis (DLA) was applied to the series and results are reported in the $\log[F(n)]$ versus $\log(n)$ plot of Fig. 8.18b. Linear interpolation yields a slope $\alpha = 1.005$, indicating that the data sequence may contain components of $1/f$-noise.

The similarity between the compositional series across the oscillatory zoning of plagioclase crystals and the numerical system indicates that they can be regarded as the product of the evolution of a dynamical system at the edge of chaos.

As we have discussed several times in the previous chapters, magma mixing processes behave as complex systems constituted by the occurrence of regular islands immersed in chaotic regions. In these two regions chemical exchanges are extremely different. Chaotic regions are characterized by faster and much more efficient mass transfer compared to regular islands.

Conceptually, this topological configuration of the mixing system has two relevant contact points with the logistic map that we have used here. Magma mixing and the logistic map are both constituted by chaotic regions and regular (periodic) regions.

We have seen in the previous section that the presence of these two regions in the mixing system can have a profound impact upon the crystallization of mineral phases. In the case of zoned pyroxenes this was reflected in the generation of different populations of crystals with different

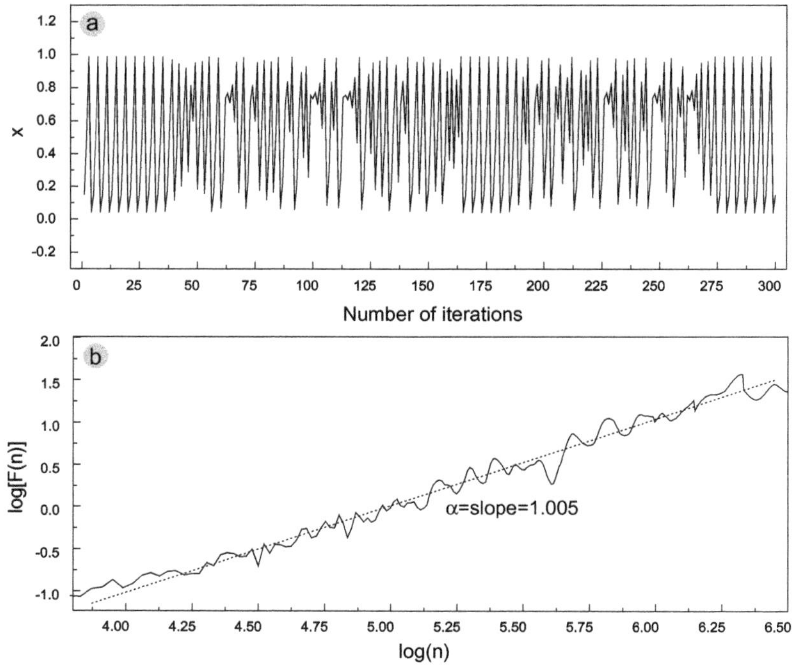

Fig. 8.18 **a** Sequence of X values deriving from the iteration of the logistic map for $r = 3.9601$; **b** variation of $\log[F(n)]$ versus $\log(n)$ for the compositional series shown in **a**. The scaling exponent α is the slope of the linear fitting of data

zoning patterns. However, the case of oscillatory zoning in the studied plagioclases is different because here they all share the common feature of displaying the fingerprint of self-organized criticality (i.e. $1/f$-noise components in the compositional patterns), defining a single crystal population.

Regular and chaotic regions might be stable in the mixing system provided that the structure of the flow field does not change in time. On the contrary, if the system configuration changes (for example due to changes in the geometry of the reservoir in which the process develops) these two regions can be continuously destroyed and generated in response to such changes.

Geometrical changes in a magmatic system can be caused by the action of external and internal triggering factors. As for the latter, crystallization is perhaps one of the most critical. Crystallization produces an increasing amount of the solid fraction in the system and this can perturb the existing flow fields forcing the system towards a new state. As crystallization is generally a continuous process, also the transition to new fluid-dynamic states can be regarded as continuous. External factors, such as changes in the aspect ratio of the magmatic reservoir due, for example, to magma migration from the magma chamber to different crustal levels, can superimpose to internal factor, strongly amplifying the perturbation of flow fields and the formation of new topological configurations of the flow. In this context, plagioclase crystals crystallize in a complex dynamical system which is continuously changing in space and time. In particular, they can be caught up into chaotic region at a certain time and immediately after they can be moved by flow fields into regular islands. This can happen many times during their crystallization.

As an example, consider the compositional profile shown in Fig. 8.13a. According to our discussion it can be said that, given the rather constant amount of Mol.% An, during the first steps of crystallization (0–25 μm) the crystal was trapped in a regular island. Successively it was transported by flow fields into a chaotic region, as suggested by the presence of a spike in the Mol.% An at ca. (45 μm). As the process progresses in time, the crystal was trapped again in a regular island developing a nearly constant variation of Mol.%

An (up to ca. (100 μm), and so forth. Similar discussions can be made for the other plagioclase crystals.

In this chapter we have attempted to provide an explanation of the complexity of mineral zoning in those magmatic systems that experienced magma mixing. Mineral zoning, however, can be generated by multiple processes possibly overlapping to the disequilibrium induce by mixing. Pressure variations due to magma ascent to the Earth surface or changes in the partition coefficients during crystallization are also factors to be considered. As for pressure, however, being it a vectorial quantity it cannot be advected by flow fields and, therefore, it should have an impact on the entire population of crystals. This means that all crystals should display, for example, one of more well identified resorption surfaces due to pressure drops during magma ascent. When mineral phases, such as those discussed here, show extremely variable zoning patterns at very short length scale it is improbable that pressure was the dominant factor. Nevertheless, caution is a must and detailed investigations should be carried out on a case-by-case basis trying to merge all the sources of information in order to draw the most complete picture about the magmatic system.

References

Anderson A Y, Anderson A T Jr (1984) Probable relations between plagioclase zoning and magma dynamics, Fuego Volcano, Guatemala. Am Mineral 69:660–676

Bak P (1990) Self-organized criticality. Physica A 163:403–409

Bak P, Tang C (1989) Earthquakes as a Self-Organized Critical Phenomenon. J Geophys Res-Solid Earth and Planets 94(B11):15635–15637

Bak P, Tang C, Wiesenfeld K (1987) Self-organized criticality: an explanation of the 1/f noise. Phys Rev Lett 59(4):381–384. http://www.ncbi.nlm.nih.gov/pubmed/10035754

Bak P, Tang C, Wiesenfeld K (1988) Self-organized criticality. Phys Rev A 38(1):364–374. http://www.ncbi.nlm.nih.gov/pubmed/9900174

Blundy JD, Shimizu N (1991) Trace-element evidence for plagioclase recycling in calc-alkaline magmas. Earth Planet Sci Lett 102(2):178–197

Couch S, Sparks RSJ, Carroll MR (2001) Mineral disequilibrium in lavas explained by convective self-mixing in open magma chambers. Nature 411(6841):1037–1039. issn: 00280836. https://doi.org/10.1038/35082540

Davidson JP, Tepley 3rd FJ (1997) Recharge in volcanic systems: evidence from isotope profiles of phenocrysts. Science 275(5301):826–829

Dobosi G, Fodor RV (1992) Magma fractionation, replenishment, and mixing as inferred from green-core clinopyroxenes in Pliocene basanite, southern Slovakia. LITHOS 28(2):133–150. issn: 00244937. https://doi.org/10.1016/0024-4937(92)90028-W

Ginibre C, Kronz A, Worner G (2002) High-resolution quantitative imaging of plagioclase composition using accumulated backscattered electron images: new constraints on oscillatory zoning. Contrib Mineral Petrol 142(4):436–448

Hergarten S (2002) Self-organized criticality in earth systems. Springer

Hibbard MJ (1994) Petrography To petrogenesis. Prentice Hall

Hu K et al (2001) Effect of trends on detrended fluctuation analysis. Phys Rev E Stat Nonlin Soft Matter Phys 64(1 Pt 1):11114

Kantelhardt JW et al (2001) Detecting long-range correlations with detrended fluctuation analysis. Physica A 295(3–4):441–454

Kuscu GG, Floyd PA (2001) Mineral compositional and textural evidence for magma mingling in the Saraykent volcanics. Lithos 56(2–3):207–230. issn: 00244937. https://doi.org/10.1016/S0024-4937(00)00051-7

May RM (1976) Simple mathematical models with very complicated dynamics. Nature 261:459–467

Nixon GT (1988) Petrology of the younger andesites and dacites of Iztacc huatl Volcano, Mexico: I. Disequilibrium phenocryst assemblages as indicators of magma chamber processes. J Petrol 29(2):213–264. issn: 0022-3530. https://doi.org/10.1093/petrology/29.2.213

Ortoleva PJ (1990) Role of attachment kinetic feedback in the oscillatory zoning of crystals grown from melts. Earth Sci Rev 29(1–4):3–8. issn: 00128252. https://doi.org/10.1016/0012-8252(0)90023-O

Pearce TH, Kolisnik AM (1990) Observations of plagioclase zoning using interference imaging. Earth-Sci Rev 29(1–4):9–26

Peng CK et al (1994) Mosaic organization of DNA nucleotides. Phys Rev E, Statistical physics, plasmas, fluids, and related interdisciplinary topics 49(2):1685–1689. http://www.ncbi.nlm.nih.gov/pubmed/9961383

Perugini D et al (2003) The role of chaotic dynamics and flow fields in the development of disequilibrium textures in volcanic rocks. J Petrol 44(4):733–756

Perugini D, Poli G, Valentini L (2005) Strange attractors in plagioclase oscillatory zoning: petrological implications. Contrib Mineral Petrol 149(4):482–497

Perugini D, Little M, Poli G (2006) Time series to petrogenesis: analysis of oscillatory zoning patterns in plagioclase crystals from lava flows. Periodico di Mineralogia 75(2–3):263–276. issn: 03698963 (ISSN)

Rutherford MJ, Hill PM (1993) Magma ascent rates from amphibole breakdown: an experimental study applied to the 1980–1986 Mount St. Helens eruptions. J

Geophy Res Solid Earth 98(B11):19667–19685. issn: 01480227. https://doi.org/10.1029/93JB01613

Shimizu N (1990) The oscillatory trace element zoning of augite phenocrysts. Earth Sci Rev 29(1–4):27–37. issn: 00128252. https://doi.org/10.1016/0012-8252(0)90025-Q

Słaby E et al (2008) K-feldspar phenocrysts in microgranular magmatic enclaves: a cathodoluminescence and geochemical study of crystal growth as a marker of magma mingling. Lithos 105:85–97

Slaby E et al (2011) Chaotic three-dimensional distribution of Ba, Rb, and Sr in feldspar megacrysts grown in an open magmatic system. Contrib Mineral Petrol 162(5):909–927. issn: 00107999. https://doi.org/10.1007/s00410-011-0631-6

Stamatelopoulou-Seymour K et al (1990) The record of magma chamber processes in plagioclase phenocrysts at Thera Volcano, Aegean Volcanic Arc, Greece. Contrib Mineral Petrol 104(1):73–84. issn: 00107999. https://doi.org/10.1007/BF00310647

Stewart ML, Fowler AD (2001) The nature occurence of discrete zoning in plagioclase from recently erupted andesitic volcanic rocks, Montserrat. J Volcanol Geotherm Res 106(3–4):243–253. https://doi.org/10.1016/S0377-0273(00)00240-7

Thompson RN (1974) Some high-pressure pyroxenes. Mineralog Ma 39(307):768–787. issn: 0026-461X. https://doi.org/10.1180/minmag.1974.039.307.04. https://www.cambridge.org/core/journals/mineralogical-magazine/article/abs/some-highpressure-pyroxenes/E84674A8B9978C7F26E50ED1947B71AB

Turcotte DL (1992) Fractals and chaos in geology and geophysics. Cambridge University Press

Umino S, Horio A (1998) Multistage magma mixing revealed in phenocryst zoning of the Yunokuchi Pumice Akagi Volcano, Japan. J Petrol 39(1):101–124. issn: 0022-3530. https://doi.org/10.1093/petroj/39.1.101. https://academic.oup.com/petrology/article-lookup/doi/10.1093/petroj/39.1.101

Wallace GS, Bergantz GW (2002) Wavelet-based correlation (WBC) of zoned crystal populations and magma mixing. Earth Planet Sci Lett 202(1):133–145. issn: 0012821X. https://doi.org/10.1016/S0012-821X(02)00762-8

Wallace PJ, Carmichael ISE (1994) Petrology of VolcánTequila, Jalisco, Mexico: disequilibrium phenocryst assemblages and evolution of the subvolcanic magma system. Contrib Mineral Petrol 117(4):345–361. issn: 00107999. https://doi.org/10.1007/BF00307270. https://link.springer.com/article/10.1007/BF00307270

Clues on the Sampling of Mixed Igneous Bodies

9

The only way of discovering the limits of the possible is to venture a little way past them into the impossible.

Arthur C. Clarke

Abstract

Magma mixing involves many variables that are non-linearly coupled and this produces magmatic bodies showing a strong geochemical heterogeneity. An interesting question arises concerning the potential loss of information due to the sampling of these complex systems for petrological studies. In particular, commonly only small portions of the magma body are available for sampling. It is therefore important to address whether the sampling can be regarded representative of the whole magmatic system. Although this is a key point in evaluating the reliability of petrological hypotheses, it is rarely addressed or even mentioned. In this chapter we attempt to address theoretically the problem of sampling of mixed igneous bodies.

9.1 Introduction

In the previous chapters we have become familiar with the fact that magma mixing involves many variables that are non-linearly coupled. This happens because the process is governed by chaotic dynamics, thus strongly amplifying the geochemical heterogeneity of igneous bodies. An interesting question arises concerning the potential loss of information due to the sampling of these complex systems for petrological studies. In particular, commonly only small portions of the magma body are available for sampling. It is therefore important to address whether the sampling can be regarded representative of the whole magmatic system. Although this is a key point in evaluating the reliability of petrological hypotheses, it is rarely addressed or even mentioned.

Instead of putting our head in the sand and proceed with our petrological and volcanological interpretations, here we attempt to address theoretically the problem of sampling of mixed bodies.

9.2 Influence of Terrain Morphology on Sampling

The $3D$ compositional field obtained with the ABC-map (Chap. 5) combining advection (stretching and folding) and diffusion will be used as an example of an igneous body generated by magma mixing. The system contains all the basic ingredients for magma mixing, including the presence of chaotic regions in which the

D. Perugini, *The Mixing of Magmas*, Advances in Volcanology,
https://doi.org/10.1007/978-3-030-81811-1_9

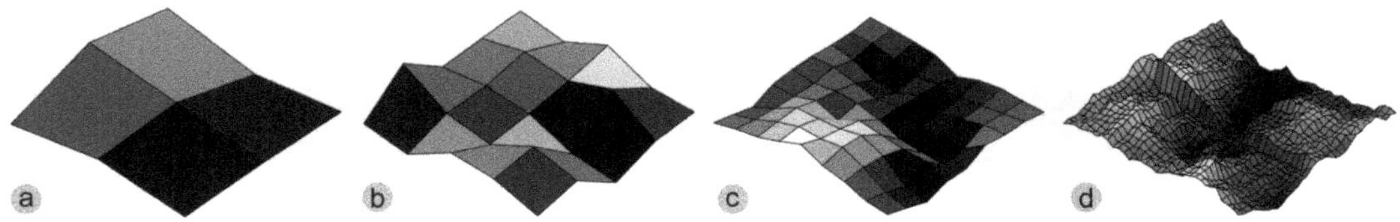

Fig. 9.1 Example of Random Mid-point Displacement Method (RMDM) applied to a plane to produce a fractal terrain

magmas can mix efficiently and regular regions where magmas roughly maintain their identity for a long time.

In the rest of this chapter, we will consider the initial fraction of mafic magma as a reference quantity to be monitored. This is taken as a proxy to evaluate the robustness of sampling.

Given the $3D$ distribution of concentrations in the magmatic mass, we can attempt estimating the uncertainty in the determination of the mafic magma fraction taking into account two of the most commonly encountered limits while sampling natural outcrops. Assuming that the simulated $3D$ body is representative of a magmatic mass produced by magma mixing, we evaluate the impact of both terrain morphology and vegetation cover on sampling.

As we have seen in Chap. 7, the patterns produced by chaotic mixing of magmas are fractals, meaning that they can be found in the magmatic mass from the metre to the micrometre length scale. The ABC map used here generates fractal compositional patterns similar to those occurring in natural systems. It appears natural, therefore, to consider it as a representative magma mixing template.

In this section we focus on the influence of terrain morphology on sampling, leaving the discussion about vegetation cover for the next section.

Several studies have shown that the Earth topography can be suitably reproduced using artificial fractal terrains (e.g. Mandelbrot 1982; Mark and Aronson 1984; Lea Cox and Wang 1993). The terrain complexity is linked to its fractal dimension (D_{land}): the larger the fractal dimension, the larger is the complexity of the terrain.

One of the most common methods used to reproduce terrains with different fractal dimension is the so-called "Random Mid-point Displacement Method" (RMDM; e.g. Mandelbrot 1982; Voss 1988). This method allows us to start with a simple geometry, such as a line or a plane, and iteratively add random features.

The simplest use of the RMDM is the generation of irregular (fractal) profiles. In this case, the starting configuration is a line. With the RMDM we displace vertically its middle point choosing a random value. This procedure is iterated many times to the middle point of each resulting segment. The shape emerging after some steps of the process produces a fractal profile (e.g. Mandelbrot 1982). The same process can be utilized for a plane; the plane is divided into four parts and a vertex is added in the middle (Fig. 9.1a). This vertex is displaced randomly along the vertical direction by a quantity δ. This process is repeated for each smaller square several times (Fig. 9.1b–d). As for the choice of the random values (δ), several works indicated that a process known as the "fractional Brownian motion" (fBm) is suitable to generate random values that are consistent with the supposed randomness of natural processes, including terrain forming phenomena (e.g. Mandelbrot 1982; Turcotte 1992). In order to generate fBm, the random number δ_i must be generated with a Gaussian distribution (mean $\mu = 0$ and standard deviation $\sigma = 1$) and, in the ith iteration, δ_i has to be modified according to the following equation:

$$\sigma_i^2 = \frac{1}{2^{2H(i+1)}}\sigma^2 \tag{9.1}$$

where H is the Hurst exponent $0 \leq H \leq 1$ (e.g., Mandelbrot 1982; Turcotte 1992). The fractal dimension D_{land} of this surface is obtained as $D_{\text{land}} = 3 - H$. As an example, Fig. 9.1d was generated by using $H = 0.5$ and, therefore, its fractal dimension is $D_{\text{land}} = 2.5$.

Artificial terrains with a fractal dimension (D_{land}) varying from 2.26 to 2.84 were generated with the RMDM. Figure 9.2 shows three repre-

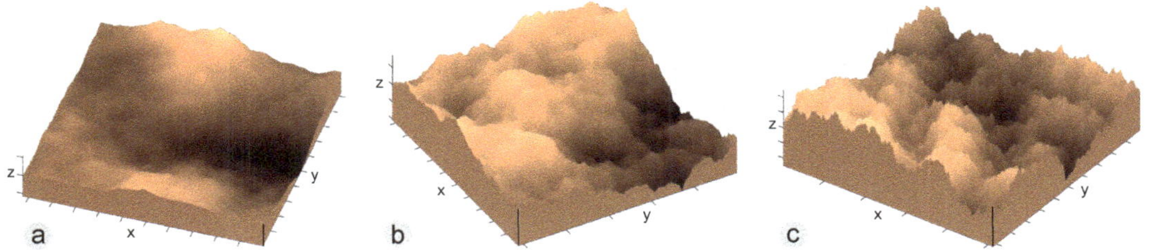

Fig. 9.2 Representative terrains with different D_{land} generated by the RMDM and used as reference for the sampling of the $3D$ concentration field (**a**: $D_{\text{land}} = 2.26$; **b**: $D_{\text{land}} = 2.58$; **c**: $D_{\text{land}} = 2.84$)

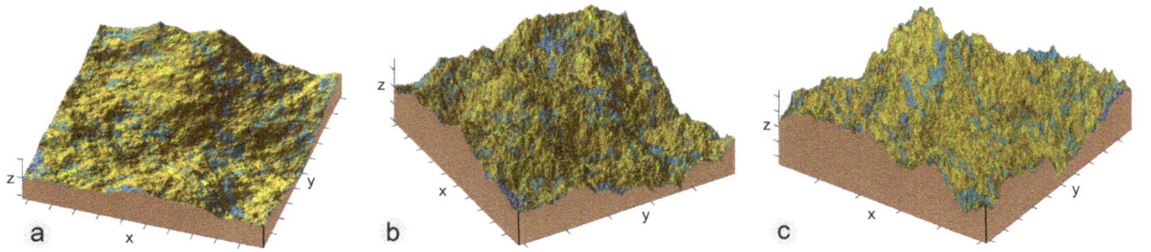

Fig. 9.3 Examples of concentration fields mapped onto the fractal terrains reported in Fig. 9.2

sentative artificial terrains with increasing fractal dimension and, therefore, increasing roughness.

Each terrain has been introduced within the $3D$ volume of concentrations produced by the chaotic mixing process (i.e. the ABC-map; Chap. 4), and concentration values corresponding to the x, y and z coordinate of the terrain were reported on the surface (Fig. 9.3). The concentration distribution on $2D$ slices similar to those shown in Fig. 4.20 (Chap. 4) were also considered. The concentrations cropping out on the terrains were sampled and utilized to derive the fraction of mafic magma participating to the mixing process. We recall that, as shown in Chap. 4, this quantity can be estimated using compositional histograms, In particular, the modal value of the histogram identifies the hybrid composition that the system would attain if the mixing was complete. For the $3D$ mixing simulations the fractions of the felsic and mafic magmas are 0.7 and 0.3, respectively. Considering that the mafic and felsic magmas have concentrations equal to zero and one, the theoretical hybrid has concentration 0.7 [i.e. $C_H = 1.0 \cdot 0.70 + 0 \cdot (1 - 0.70)$], which corresponds to the maximum value on the histogram. As we have discussed in the previous chapters, the hybrid composition of the mixing system is an important parameter. It allows us, using the two end-member mixing equation, to estimate the initial fractions of end-member magmas by knowing their initial compositions.

Figure 9.4a shows histograms of concentrations fitted by a Gaussian function for the whole $3D$ volume and a $2D$ plane, and two fractal terrains with different fractal dimension. In the graph, concentration frequency values were scaled between zero and unity to compare the different distributions. The concentration variability on the $2D$ surface and the terrain with $D_{\text{land}} = 2.26$ shows the most significant deviation from the whole $3D$ volume. As the fractal dimension of the terrain increases (e.g. $D_{\text{land}} = 2.84$), the spectrum of concentrations tends to overlap with the distribution of concentrations of the whole $3D$ volume.

The plot of Fig. 9.4b displays that the accuracy in estimating the initial fraction of the mafic magma increases as D_{land} of the terrain increases. In particular, the largest errors (of the order of 40%) are recorded considering the concentration distributions on $2D$ planes. As the fractal dimen-

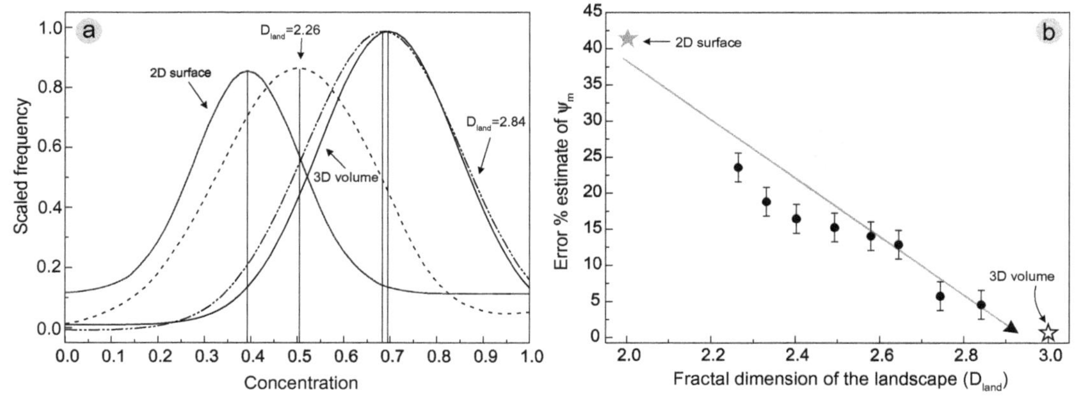

Fig. 9.4 **a** Concentration histograms fitted by a Gaussian function for the whole 3D volume, a 2D plane, and two fractal terrains with different fractal dimension; projection of maximum values of functions on the x-axis is also reported; **b** variation of uncertainty in the estimate of the mafic magma fraction (expressed as error % with respect to the actual value) against fractal dimension of the terrains. Modified from Perugini et al. (2007)

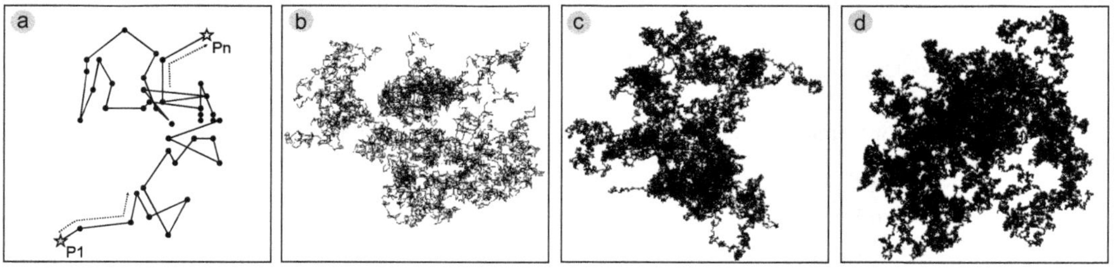

Fig. 9.5 **a** Brownian motion in two dimensions; **b–d** vegetation patches with different areal extension used to test the influence of vegetation cover on sampling

sion of the terrain increases, errors drop quickly reaching values of ca. 5% for terrains with fractal dimension of 2.76–2.84. This indicates that the roughness of the terrain, quantified by its fractal dimension, strongly influences the robustness of the sampling of mixed igneous bodies.

9.3 Influence of Vegetation Cover on Sampling

In the discussion above, we considered that the whole surface was available for sampling; however, this is not common because several factors typically limit sampling of such large outcrops. In particular, very often outcrop portions are not reachable because they are pervaded by vegetation. In order to consider this limiting circumstance and its impact upon the quality of sampling, the terrains previously generated were covered with variable percentages (from 10 to 90%) of vegetation patches.

In particular, the vegetation cover was produced with a random Brownian motion process in 2D. This allows us to generate vegetation patches which are topologically equivalent to natural vegetation (e.g. Mandelbrot 1982). Brownian motion in 2D occurs by moving a point randomly on a 2D surface. In detail, the steps moved by the point in the x and y direction are random, normally distributed, values (e.g. Mandelbrot 1982).

Some initial steps of Brownian motion are given in Fig. 9.5a, where a particle initially positioned at point $P1$ moves erratically across the plane. This approach was utilized to generate patches of random vegetation covering different percentages of the fractal terrains used previously. Some examples of vegetation patches are shown in Fig. 9.5b–d

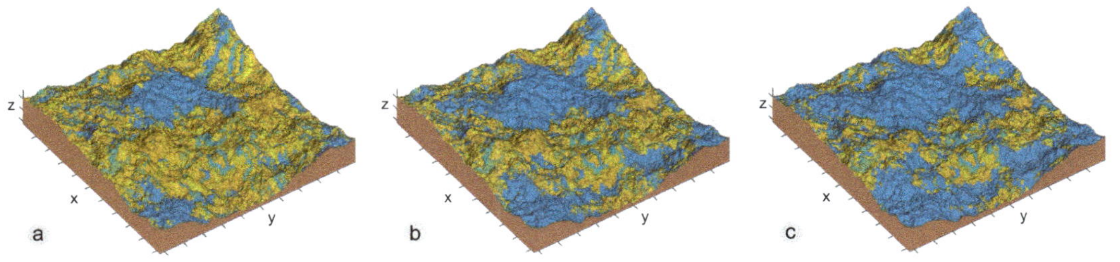

Fig. 9.6 Fractal terrain shown in Fig. 9.3a covered by different percentages of vegetation (blue patches)

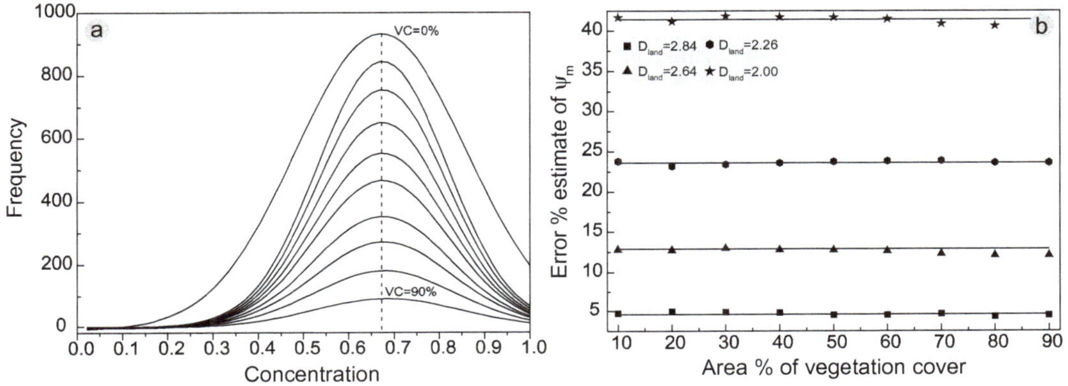

Fig. 9.7 **a** Variation of concentration histogram fitted by a Gaussian function for different percentages of vegetation cover for the fractal terrain with $D_{\text{land}} = 2.84$; **b** variation of accuracy in the estimate of mafic magma fraction as a function of vegetation cover for terrains with different D_{land}

where, in order to generate different percentages of vegetation cover, the Brownian motion was allowed to progress at different times.

Figure 9.6 shows the fractal terrain reported in Fig. 9.3a and covered by an increasing percentage of vegetation. Histograms of concentration have been constructed considering only those concentrations on outcrop portions of the fractal terrains that were not hidden below the vegetation cover. Figure 9.7a shows the variation of the concentration histograms for different percentages of vegetation for the fractal terrain with $D_{\text{land}} = 2.84$. The graph displays that the modal value of the histogram does not change by increasing the vegetation coverage. The plot of Fig. 9.7b shows the variation of the accuracy in the estimate of the mafic magma fraction as a function of the percentage of vegetation cover for terrains with different irregularity. It is shown that the modal value of the concentration histogram for all fractal terrains is not influenced by the amount of vegetation coverage. This means that the uncertainty in the sampling of a mixed igneous mass remains constant as the amount of available outcrop decreases due to vegetation cover, no matter how rough the terrain is.

Recapitulating, the sampling of small parts (as small as 10% of the entire igneous body) of very rough outcrops will always provide higher quality information relative to sampling 100% of flat or nearly flat outcrops.

9.4 Concluding Remarks

We have seen that sampling of mixed igneous bodies might not be a trivial task and that the quality of sampling is strongly dependant on roughness of the terrain on which such an igneous body crops out. The robustness of the sampling can be efficiently improved by sampling outcrop portions displaying the largest roughness, and excluding flat or nearly flat areas. We have also seen that vegetation coverage is not likely to have an impact upon the quality of sampling, and that small-scale but very irregular outcrops contain the essential information and should be preferred to large and flat or nearly flat ones.

References

Lea Cox B, Wang JSY (1993) Fractal surfaces: measurement and applications in the earth sciences. Fractals 01(01):87–115. ISSN: 0218-348X. https://doi.org/10.1142/s0218348x93000125

Mandelbrot BB (1982) The fractal geometry of nature. W.H. Freeman

Mark DM, Aronson PB (1984) Scale-dependent fractal dimensions of topographic surfaces: an empirical investigation, with applications in geomorphology and computer mapping. J Int Ass Math Geol 16(7):671–683. ISSN: 00205958. https://doi.org/10.1007/BF01033029

Perugini D, Petrelli M, Poli G (2007) Influence of landscape morphology and vegetation cover on the sampling of mixed plutonic bodies (English). Mineral Petrol 90(1–2):1–17. ISSN: 09300708. https://doi.org/10.1007/s00710-006-0173-1

Turcotte DL (1992) Fractals and chaos in geology and geophysics. Cambridge University Press, Cambridge

Voss RF (1988) The science of fractal images. In: Peitgen HO, Saupe D (eds.) Chap. Fractals i. Springer, Berlin, pp 21–70

Part IV

Magma Mixing: A Volcanological Tool

Magma Mixing: The Trigger for Explosive Volcanic Eruptions

10

There are two possible outcomes: if the result confirms the hypothesis, then you've made a measurement. If the result is contrary to the hypothesis, then you've made a discovery.

Enrico Fermi

Abstract

Volcanic eruptions are potentially catastrophic phenomena that can have a huge impact on society and the environment. Understanding the causes, dynamics and timescales of eruptions is of greatest importance for mitigating the medium- to large-scale impact of these natural events. In this chapter we show that the most explosive volcanic eruptions occurred on planet Earth are associated with magma mixing processes. Discriminating whether the mixing process was the cause or the effect of the eruptions is not easy, although there is evidence that mixing might have been the triggering factor. In any case, being magma mixing a common factor it is not wise to ignore its occurrence as it may help in shedding light on eruption dynamics. This will be the focus of this chapter.

10.1 Magma Mixing and Volcanic Eruptions

Volcanic eruptions are potentially catastrophic phenomena that can have a huge impact on society and the environment. Understanding the causes, dynamics and timescales of eruptions is of greatest importance for mitigating the medium- to large-scale impact of these natural events.

Volcanic explosions are the "end-product" of the petrological processes that occurred in the magmatic mass since its formation in the source region to the migration to the Earth's surface. During its lifecycle the composition of a magma body can experience modifications caused by many different and overlapping processes such as incorporation and assimilation of country rocks, fractional crystallization, and magma mixing.

Discerning the relative importance of each of these processes is often a challenging task and petrologists generally do their best using petrological and geochemical modeling. This, combined with experimental petrology usually furnishes a first sketch of the complexity of the evolutionary processes that produced the erupted magmas.

D. Perugini, *The Mixing of Magmas*, Advances in Volcanology,
https://doi.org/10.1007/978-3-030-81811-1_10

For the petrologist this is generally enough. However, petrological processes sometimes can provide the opportunity to penetrate in greater detail into the complex life of magmas, from their birth deep into the Earth's interior to their final destination onto the Earth's surface as lava flows or pyroclastic deposits.

Finding a common denominator linking petrology and volcanic eruptions would equip us with a formidable tool to track the evolution of volcanic systems, greatly simplifying our goal of better understanding volcanic eruptions and their timescales.

Table 10.1 shows some of the major eruptions that occurred on planet Earth in the last two centuries with VEI (Volcanic Explosivity Index) equal to or larger than three. As the author was born in Italy also the catastrophic eruption of Vesuvius that occurred in A.D. 79 is included. A bibliographic research on articles published on these eruptions unequivocally indicates that all of them share a common feature: they are all associated with magma mixing processes. There is no doubt that the eruptions listed in the table represent only a handful of eruptions. Nevertheless, I challenge the readers to perform a more exhaustive research, possibly going back in time considering older eruptions. I would be very surprised if they can estimate a percentage of explosive eruptions associated with magma mixing lower than 90%.

Discriminating whether the mixing process was the cause or the effect of the eruptions is not easy, although several authors argued that mixing was the triggering factor (e.g. Druitt et al. 2012; Leonard et al. 2002; Martin et al. 2008; Kent et al. 2010). In any case, being magma mixing a common factor it is not wise to ignore its occurrence as it may help in shedding light on both the dynamics and timescales of eruptions. This will be the focus of this and the next chapter.

10.2 Dynamics and Time Evolution of Plumbing Systems

10.2.1 Deciphering Magma Chamber Evolution and Estimation of Eruptive Activity Using Magma Mixing

In Chap. 7 we have discussed the occurrence of magmatic enclaves in the Pietre Cotte lava flow, erupted on the Island of Vulcano (Aeolian Islands, Italy) in A.D. 1739. Here latitic enclaves are dispersed in an obsidian rhyolitic mass (Fig. 7.3).

We analysed using fractal techniques both the size distribution and the morphology of enclaves and hypothesized that they might represent the remnants of viscous fingering structures formed by the injection of the latitic magma into a rhyolitic magma chamber in the plumbing system of the volcano. Enclaves, therefore, are considered to preserve in their shapes information about magma chamber dynamics. In particular, enclave morphology is considered to reflect the viscosity ratio (V_R) between the mafic and the felsic magma.

Viscosity ratios between the two magmas were estimated using the empirical relationship given in Eq. 6.1 and results indicate values of viscosity ratios (V_R) in the range 2.3–5.5, with most values clustered around V_R=2.6 (Fig. 7.4).

In the attempt to provide an explanation for the estimated variability of viscosity ratios, we can evaluate different hypotheses, using thermodynamical and rheological modelling. Values of the specific model variables are as follows: $T_m = 1283$ K, $T_f = 1223$ K, $\rho_m = 2587$ kg m^3, $\rho_f = 2270$ kg m^3, $C_m = 1285$ J K^{-1} kg^{-1}, $C_f = 1285$ J K^{-1} kg^{-1}. In the modelling we assume that the average composition of the latitic enclaves and the rhyolite represent the composition of the initial end-member magmas.

Table 10.1 Some of the major eruptions that occurred on planet Earth in the last two centuries with VEI equal to or larger than three. References as follows: a: Sparks et al. (1977); b: Pallister et al. (1992); c: Murphy et al. (1998); d: Cioni et al. (1995); e: Leonard et al. (2002); f: Sigmundsson et al. (2010); g: Clynne (1915); h: Gardner et al. (1995); i: Tepley et al. (2000); j: Gerbe et al. (1992); k: Williams and Self (1983); l: Volynets et al. (1999); m: Gertisser et al. (2012); n: Reagan (2003); o: Robin et al. (1991); p: Davydova et al. (2018); q: Self and Rampino (2015); r: Kratzmann et al. (1991)

Volcano	VEI	Year (AD)	Country	Magma A	Magma B	Ref.
Askja	5	1875	Iceland	basalt	rhyolite	a
Pinatubo	6	1991	Philippines	basalt	dacite	b
Soufriere Hills	3	1995	Montserrat (UK)	basalt	andesite	c
Vesuvius	5	79	Italy	basanite	phonolite	d
Krakatau	6	1883	Java (Indonesia)	basalt	rhyolite	e
Eyjafjall	3	2010	Iceland	basalt	dacite	f
Lassen Peak	3	1915	California (USA)	basalt-andesite	dacite	g
St. Helens	5	1980	Washington (USA)	basalt	dacite	h
El Chichon	5	1982	Mexico	basalt	trachyte	i
Gallunggung	5	1822	Java (Indonesia)	basalt	basalt-andesite	j
Santa Maria	6	1902	Guatemala	basalt	dacite	k
Ksudach	5	1907	Kamchatka (Russia)	andesite	rhyodacite	l
Tambora	7	1815	(Indonesia)	basalt	trachy-andesite	m
Novarupta	6	1912	Alaska (USA)	andesite	rhyolite	n
Fuego de Colima	5	1913	Mexico	basaltic-andesite	andesite	o
Bezymianny	5	1956	Kamchatka (Russia)	basaltic-andesite	dacite	p
Agung	5	1963	Bali (Indonesia)	basalt	andesite	q
Cerro-Hudson	5	1991	Chile	basalt	trachy-andesite	r

AlphaMELTS (Smith and Asimow 2005) was used to estimate the viscosity of the rhyolitic host mass at the liquidus temperature ($T_{fi} = 1223$ K). Its viscosity variation with temperature was modelled according to Giordano et al. (2008).

Regarding the latitic magma, its average crystal content varies in the restricted range 25–30 vol.%. Crystals do not show evidence of physical and chemical disequilibrium and are, therefore, interpreted as the crystal content of the enclaves before it came into contact with the rhyolitic one.

The Einstein-Roscoe equation expresses the influence of crystals on viscosity in magmas for crystals with aspect ratio lower than three (e.g. Vona et al. 2011; Vetere et al. 2013; Mader et al. 2013):

$$\eta_r = \frac{\eta}{\eta_{\text{melt}}} = \left(1 - \frac{\chi}{\alpha}\right)^{-2.5} \quad (10.1)$$

where η_r is relative viscosity, η_{melt} is melt viscosity, χ is crystal fraction and α is a constant (Marsh (1981), Roscoe (1952)). Marsh (1981) suggested for magmatic systems the following modified version

$$\eta = \eta_{\text{melt}} \cdot \left(1 - \frac{\chi}{0.6}\right)^{-2.5} \quad (10.2)$$

We used the model of Giordano et al. (2008) to estimate η_{melt} (i.e. the viscosity of the mafic melt) using the composition of the liquid obtained with alphaMELTS.

During the intrusion of the latitic magma into the rhyolitic mass, thermal exchanges between the two magmas must have occurred. In particular, it is likely that the mafic and felsic magmas underwent cooling and heating, respectively.

Following Folch and Marti (1998) the variation of temperature of the felsic ΔT_f and mafic ΔT_m magmas can be expressed as:

$$\Delta T_f = \frac{\varphi \rho_m C_m}{\varphi \rho_m C_m + \rho_f C_f} \left(T_{mi} - T_{fi}\right) \quad (10.3)$$

$$\Delta T_m = \frac{\rho_f C_f}{\varphi \rho_m C_m + \rho_f C_f} \left(T_{fi} - T_{mi}\right) \quad (10.4)$$

where T_{mi} and T_{fi} are the initial temperatures of the mafic and the felsic magma, C_m and C_f are their relative specific heat capacities, ρ_m and ρ_f their density, and φ is the volume fraction of the mafic magma defined as $\varphi = V_{mi}/V_{fi}$ (V_{mi} and V_{fi} are the relative volumes of mafic and felsic magmas).

According to our discussion in Chap. 7, any plausible process must satisfy the requirement that the viscosity of the mafic magma always had a lower viscosity compared to the felsic magma. This is constrained by the fact that viscous fingering dynamics can develop only in these rheological conditions.

Values of intensive variables utilised in the thermodynamic modelling are: starting temperature of the simulation $T_{ms} = 1523$ K and pressure $P = 600$ bar (ca. 1.5–2.0 km depth), i.e. the estimated depth of the shallow plumbing system of Vulcano (Clocchiatti et al. 1994; Piochi et al. 2009; Zanon 2003).

The liquidus temperature of the latite estimated by the alphaMELTS simulation is $T_{liq} = 1313$ K. Cooling of the latitic melt triggers crystallization of the latite (Fig. 10.1). After 25 % crystallization (i.e. the lower limit of crystal content of the enclaves), and considering the melt composition provided by the simulation, the viscosity of the latite (Eq. 10.2) is $\log(\eta_{mi}) = 4.84$ Pas at $T_{mi} = 1283$

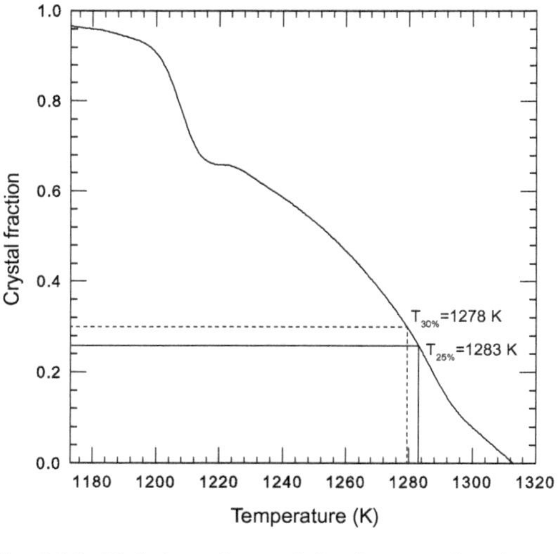

Fig. 10.1 Variation of crystal fraction as a function of temperature for the cooling of the latitic magma

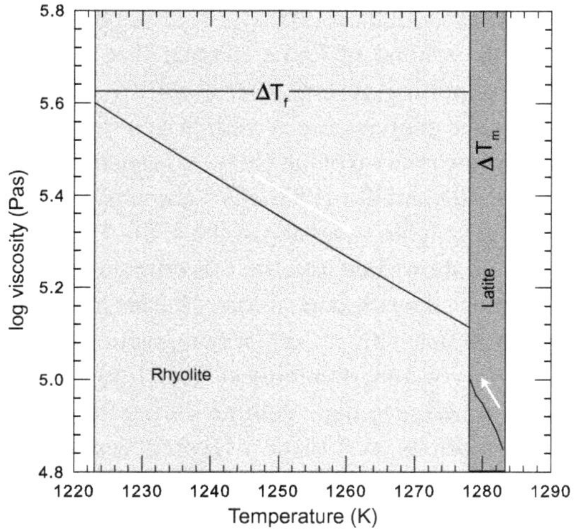

Fig. 10.2 Variation of the viscosity of the felsic and mafic magmas. White arrows indicate the relative direction of variation of viscosity

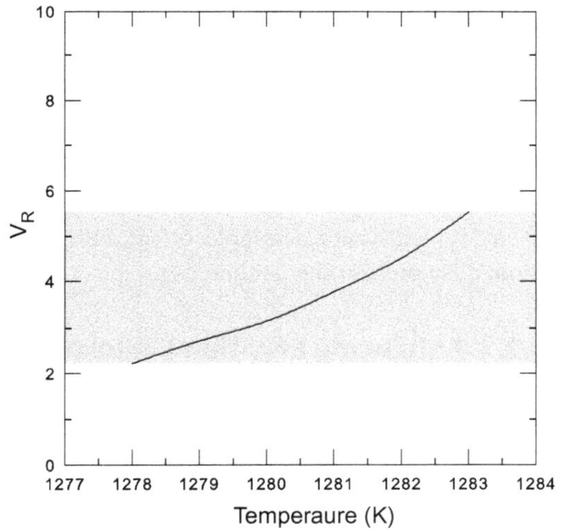

Fig. 10.3 Variation of V_R as a function of temperature. The grey area represents the range of V_R estimated from fractal analysis

K. The latitic magma, therefore, is assumed to have these values of viscosity and temperature before invading the rhyolitic magma chamber.

Equation 10.4 was applied considering the values of the variables defined above. Results indicate that the variation of temperature of the mafic magma is $\Delta T_m = 40$ K and $\Delta T_m = 60$ K for volume fractions of the mafic magma (φ) in the range 0.01 and 0.43. Considering the simulation of crystallization of the latitic magma (Fig. 10.1), these agree with crystal contents larger than 60 vol. % (i.e. $\chi = 0.6$). Following Marsh (1981) this represents the threshold value above which the magma in unable to flow. However, this in contrary to field observations and analysis of enclave morphologies indicating that the mafic magma was, indeed, able to flow and invade the felsic magma originating viscous fingering structures.

Additional examination of the crystallization curve of the latitic magma (Fig. 10.1) indicates that T_m was not likely to be lower than 1278 K because this would imply crystal fractions of the latitic magma larger than 30 vol. % ($\chi > 0.3$), which is the largest crystal fraction measured on the latitic enclaves.

These considerations indicate that in order to take into account both field evidence and thermodynamic modelling, only volume fractions of the latitic magma larger than 90% appear plausible.

The variation of temperature of the felsic and mafic magmas was calculated using Eqs. 10.3 and 10.4, considering fractions of the latitic and rhyolitic magma of 0.9 and 0.1. ΔT_f and ΔT_m are 55 K and 5 K, respectively.

The viscosity variation of the two magmas in the above ranges of temperature is presented in Fig. 10.2. Given that the temperature of the rhyolitic magma remains above its liquid temperature, the viscosity variation is calculated using the model of Giordano et al. (2008). The viscosity variation of the latitic magma (η_{melt}) is calculated with Eq. 10.2 using Giordano et al. (2008) and the composition of the residual liquid obtained with the alphaMELTS simulation in the interval of crystal fractions from 0.25 to 0.30 (i.e. the maximum crystal fraction measured in the enclaves). Figure 10.2 shows that the rhyolite experiences the largest viscosity variation, moving progressively towards the viscosity of the latitic magma.

Figure 10.3 displays the variation of viscosity ratio (V_R) between the two magmas as a function of temperature in the range of viscosity variations shown in Fig. 10.2. In particular, V_R is calculated for the variation of viscosity [$\log(\eta)$] of the latitic magma in the range 4.8-5.0 and for temperatures from 1278 K to 1283 K. The viscosity of the felsic magma is taken constant at the value $\log(\eta) = 5.16$

at $T = 1278$ K. The plot (Fig. 10.3) also shows the range of V_R estimated from the fractal analysis of enclaves (Fig. 7.4) indicating a good agreement between the thermodynamic simulation considering large amount of the latitic magma and V_R values estimated from fractal analysis.

The above results indicate that the rhyolitic mass was overheated by the arrival of the latitic magma in the magma chamber. This is also supported by results presented by Clocchiatti et al. (1994) who estimated eruptive temperatures up to 1273 K. This temperature agrees well with the results presented above where the overheating of the rhyolitic melt to this temperature was caused by the arrival of about 90% of latitic magma in the plumbing system of the volcano. Overheating of the rhyolitic mass may have enhanced its mobility, favoring magma ascent and eruption.

According to the above discussion, hence, large amounts of latitic magma were injected into the sub-volcanic magma chamber before and/or during the A.D. 1739 activity of the volcanic system.

The total volume of the Pietre Cotte lava flow is estimated of the order of $2.5 \cdot 10^6$ m^3 Frazzetta and La Volpe (1991). Considering that latitic enclaves constitute 5.0–6.0 vol.% of the total volume, it can be calculated that the relative volumes of rhyolite and latite forming the lava flow is ca. $2.3 \cdot 10^6$ and $1.2 \cdot 10^5$ m^3, respectively. If we assume that the whole amount of rhyolite was erupted, we can estimate that ca. $2.0 \cdot 10^7$ m^3 of latitic magma remained in the sub-volcanic magma chamber after the eruption of the Pietre Cotte lava flow.

After a period of 150 years of minor activity, a new eruptive cycle started on the Islands of Vulcano from 2 August 1888 to 22 March 1890. This eruptive period was characterized by explosions producing pyroclastic deposits with trachy-rhyolitic and latitic compositions (Clocchiatti et al. 1994, De Astis et al. (2006)). It is interesting to note that the geochemical composition of these eruptive products encompasses the geochemical variability of Pietre Cotte rocks. Regarding the trachy-rhyolitic compositions erupted in 1888–1890, they are considered as the result of low-pressure fractional crystallization processes starting from a latitic parental magma (Clocchiatti et al. 1994, Zanon (2003)). This suggests that the large volume of latitic magma that was left in the plumbing system after the 1739 eruption might have evolved to more silica rich melts during the time interval of the 150 years separating the Pietre Cotte and the 1888–1890 eruptive activity. The stratigraphic sequence of the 1888–190 eruptive cycle shows that rhyolite was erupted first and was followed by the deposition of the least evolved compositions. This reflects a compositional zoning in the volcanic plumbing system. Emptying of such a zoned magmatic column during the 1888–1890 eruptions must have triggered widespread mixing processes, which are indeed observed in the eruptive products (Clocchiatti et al. 1994).

We have shown here that it is possible to track the dynamics and the time-evolution of shallow plumbing systems through the study of the mixing patterns preserved in the erupted rocks. We have also shown that this allows us to obtain information about volume proportions of the magmas that participated to the mixing process. The method might represent an additional volcanological tool for hazard mitigation in volcanic areas because it provides the opportunity to obtain information that are typically not available or difficult to be obtained by present-day technology.

10.2.2 Enhancing Eruption Explosivity by Magma Mixing

A further interesting example in which magma mixing can be used to track the evolution of a volcanic plumbing system is provided by the caldera-forming eruption of Sete Cidades (Island of Sao Miguel, Azores archipelago; 16 Ky B.P.).

Studied samples belong to the plinian fallout deposit in the Upper member of the Santa Barbara unit (Queiroz et al. 2015; Beier et al. 2006; Paredes-Mariño et al. 2017). Samples are trachytic pumices in which basaltic fragments with a highly variable size range are dispersed. Trachyte samples are strongly vesiculated and mostly aphyric; basaltic fragments show larger crystal contents and scarce vesiculation (Paredes-Mariño et al. 2017).

3D microCT reconstructions of two representative samples are show in Figs. 10.4 and 10.5. The

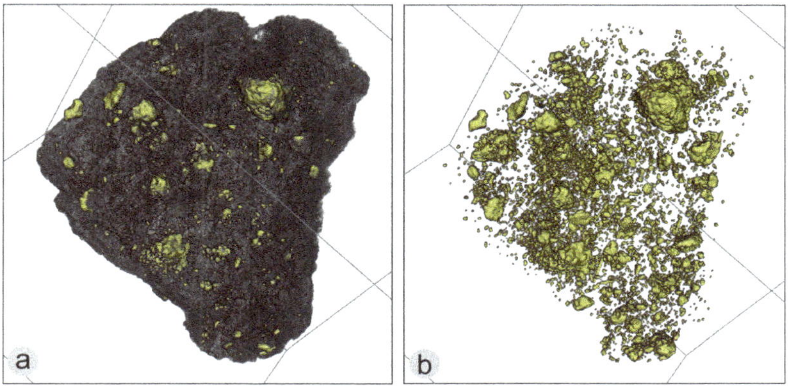

Fig. 10.4 **a** MicroCT of a sample from the Santa Barbara outcrop. Trachytic pumice and basaltic fragments are reported in the grey and yellow colour, respectively; **b** 3*D* distribution of basaltic fragments

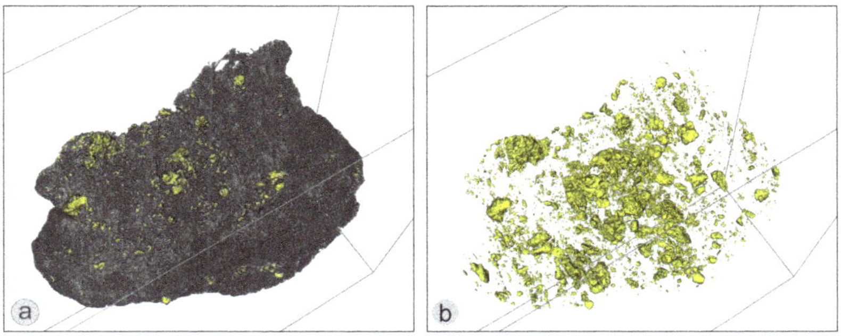

Fig. 10.5 **a** MicroCT of a sample from the Santa Barbara outcrop. Trachytic pumice and basaltic fragments are reported in the grey and yellow colour, respectively; **b** 3*D* distribution of basaltic fragments

left panels of each figure (Figs. 10.4a, 10.5a) display the samples in which both the trachyte and the basaltic fragments are present. The left panels (Figs. 10.4b, 10.5b) show the basaltic fragments after the removal of the trachytic component. The images show that the basaltic fragments are present in a wide range of sizes and appear distributed into the pores (bubbles) of the trachyte. This last feature can be clearly seen in the sample sections of Fig. 10.6. Figure 10.6a shows a bubble of the trachyte in which a basaltic fragment occurs. The fragment is characterized by intense fracturing and the resulting fragments show a great variability of fragment sizes, comparable to that of the fragments found dispersed around the perimeter of other bubbles. This can be seen in the pictures of Fig. 10.6b, c where the expansion of the bubbles appears to have dismembered larger basaltic fragments and the resulting pieces are now coating the bubble walls.

Given that the large amount and variability of basaltic fragments in the trachyte appears a key feature of the studied rocks, fractal fragmentation analysis was performed. Previously we have used this method to study the size distribution (area) of magmatic enclaves (see Chap. 7). In this case, since the volume of the fragments is directly available from microCT 3*D* reconstructions, the analysis is applied considering the volume of basaltic fragments. In the analysis, since the size comparison is performed using a volume ($V > v$), Eq. 7.1 becomes

$$N(V > v) = kv^{-D_f/3} \tag{10.5}$$

with $D_f = -3m$, where m is the slope of the linear fitting.

Figure 10.7 shows the variation of $\log[N(V > v)]$ as a function of $\log(v)$, according to Eq. 10.5. As the samples belong to the same pyroclastic deposit, the graph was constructed by merging volumes of the basaltic fragments from different samples. The graph indicates that data points are disposed along a straight line, fulfilling the requirement for a fractal-fragmented distribution (e.g. Turcotte 1992). Linear interpolation of data

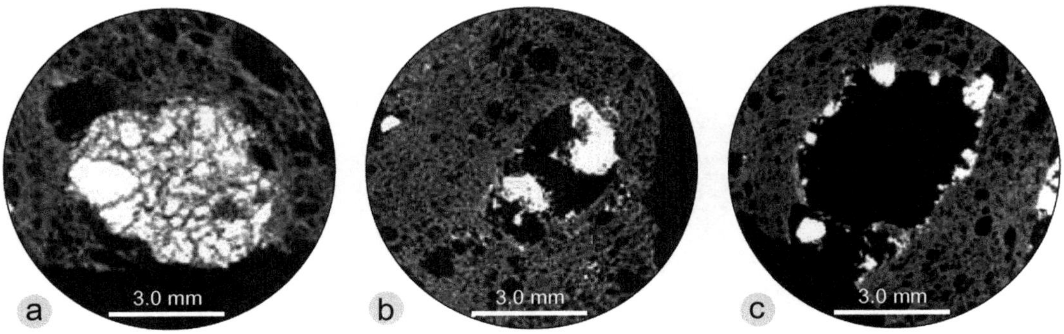

Fig. 10.6 Representative BSE sections of studied samples showing the distribution of basaltic fragments into the vesicles of the trachyte

Fig. 10.7 Variation $\log[N(V > v)]$ against $\log(v)$ representing the size distribution of basaltic fragments. Solid line is the linear fitting of the data

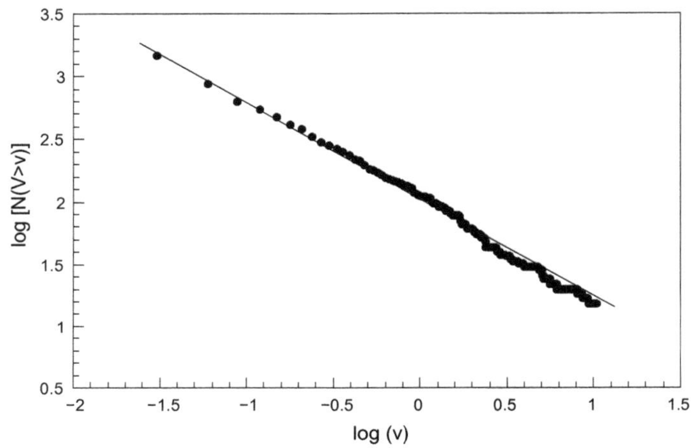

gives a slope $m = -0.858$ and, consequently, a value for fractal dimension $D_f = 2.57$.

As discussed in Chap. 2, the conceptual model used to derive Eq. 10.5 is based on the self-similar fragmentation of a mass into progressively smaller particles (Fig. 2.12). Calculation of fractal dimension of fragmentation of such a model yielded a value of $D_f = 2.601$. This value of D_f has been measured for a variety of rock types and appears to be a typical value for fragmentation of materials with solid-state rheological behaviour (e.g. Turcotte 1992).

As shown above, the analysis of the fractal size distribution of basaltic fragments gives a value of fractal dimension of fragmentation (D_f) equal to 2.57. As this value is very close to the value of D_f measured for solid-state fragmentation, the fragment size distribution of basaltic fragments can be considered the result of fragmentation of the basalt when it was in solid rheological conditions. This result is also supported by petrographic observations indicating that basaltic fragments have sharp edges and cuspate margins, a feature typically displayed by fragmentation of solid materials (Paredes-Mariño et al. 2017).

Different hypotheses can be considered for explaining the occurrence of basaltic fragments in the trachytic pumices and they are discussed in details in Paredes-Mariño et al. (2017). Among them, the most reliable one is that the basaltic magma was injected into the trachytic magma at depth, where it fragmented, and ascended together with the trachyte until the magmatic mixture exploded triggering the eruption. Robust indications for the appropriateness of this hypothesis can be found in the studied rocks and, in particular, in the petrographic features of the basaltic fragments. In particular, undercooling textures of mineral phases indicate that the basaltic magma underwent quenching (e.g. Bacon et al. 1986; Hibbard 1994) and this agrees with the thermal shock that the basalt would have suffered during injection in the trachytic magma. The rapid cooling moved the rheological behaviour of the basaltic component towards that of a solid, as supported by the value of fractal dimension fragmentation of basaltic fragments ($D_f = 2.57$).

The release of heat from the basaltic magma, possibly reinforced by the additional latent heat released during its rapid crystallization, might have triggered convection dynamics in the trachytic magma (Snyder (2000), Oldenburg et al. 1989). This might have had a double effect on the system: (i) the trachytic melt experienced a decrease in viscosity (Sparks and Marshall 1986, Perugini and Poli 2005) due to the heat released by the basaltic magma and (ii) basaltic fragments generated by the disruption of the initial basaltic body were dispersed throughout the trachytic melt. The fluidization of the trachytic magma allowed it to acquire a larger mobility possibly forcing it to ascend towards the Earth surface (Kent et al. (2010), Perugini et al. (2015)). During magma ascent, bubble nucleation and growth in the trachyte occurred. When the magma reached the fragmentation level, it exploded and the eruption developed.

Although this might appear a viable hypothesis that can account for the features observed in the studied rocks, some issues still require clarification. The problem of the occurrence of basaltic fragments coating the interior of bubbles in the trachytic pyroclasts is perhaps the most intriguing one. Possible explanations must be searched in the time interval between the nucleation and growth of bubbles and the volcanic explosion, i.e. the time at which the magmatic history of the system ended and the textural and structural features of the studied rocks were frozen in time. A hypothesis that might explain this peculiar feature is the heterogeneous nucleation of bubbles in the trachyte due to the presence of basaltic solid fragments.

The presence of solid phases in magmas can strongly boost the formation of bubbles. In particular, several works based both on natural observations and decompression experiments (e.g. Hurwitz and Navon 1994; Mangan and Sisson 2000; Gardner and Heene 2004; Gardner 2007) have shown that, for example, heterogeneous nucleation of bubbles on magnetite grains is a common phenomenon. This happen because this mineral provides the most energetically favourable surface for heterogeneous nucleation of bubbles. These results were corroborated by the work of Gualda and Ghiorso (2007) who extended the process of heterogeneous nucleation of bubbles also to silicate mineral-bubble aggregates. This indicates that, potentially, crystal-bearing xenoliths (represented in our case by the basaltic fragments in the trachyte) can act as favourable sites for heterogeneous nucleation of bubbles. The presence of basaltic fragments coating the interior of the bubbles (Fig. 10.6) is a consequence of this process. Indeed, as shown in Fig. 10.6, single basaltic fragments appear intensely fractured in minute particles. Nucleation and growth of bubbles around them can pull them apart and, consequently, those smaller basaltic particles can remain "glued" to the inner walls of the growing bubbles. The process occurred in the whole trachytic magmatic mass until the life of the magmatic system was terminated by the eruption.

Resuming, the presence of basaltic fragments might have acted as energetically favourable sites to trigger bubble nucleation and growth in the trachytic melt. This would have facilitated the gas exsolution from the trachyte, possibly enhancing the explosivity of the eruption once the magmatic mass reached the fragmentation level. Figure 10.8 reports schematically the hypothesized evolution of the system from the injection of the basaltic magma into the trachytic melt at depth to eruption.

10.3 Using Mixing Patterns to Infer the Dynamics of Explosive Eruptions

In the previous sections we discussed the possible use of enclaves/fragments generated by magma mixing as markers of the dynamics occurring in

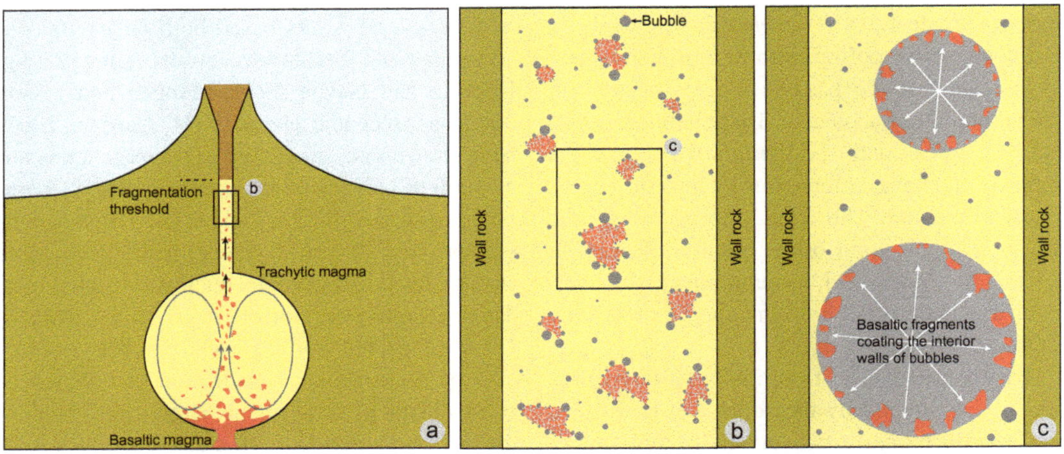

Fig. 10.8 Schematic representation of the evolution of the volcanic system since the injection of the basaltic magma into the trachytic chamber to the fragmentation level in the volcanic conduit. **a** Injection of the basaltic magma in the trachytic chamber and generation of thermodynamical instability; **b** zoomed-in view of the system during magma migration in the conduit. The basaltic fragments behave as sites for bubble nucleation in the trachytic melt; **c** growth of bubbles around the basaltic fragments provoked the detachment of smaller pieces of basaltic rock that remained attached to the inner walls of the bubbles

the plumbing system. In this section we will focus on the potential applicability of magma mixing patterns to deciphering the fluid dynamics of explosive eruptions.

As an example, we consider again the mixed pyroclasts constituting the Pollara deposit on the Island of Salina (Aeolian Islands, Italy). Pictures of representative samples are reported in Fig. 7.20, where we have shown that the gray value of the images correlates with the chemical composition of the glass. We have also shown that the maximum frequency of the compositional histogram corresponds to the hybrid composition that the system would attain in the case of complete mixing.

Building on this, we now attempt to estimate the Reynolds number (Re) that characterized the mixing process in the conduit.

On the basis of fluid dynamic experiments of turbulent jet mixing Catrakis and Dimotakis (1998) have derived an empirical relationship between the concentration patterns and the relative Reynolds numbers (Re). They found a relation between Re and the perimeter of contours corresponding to the modal concentration of mixing patterns (P_{tot}) normalized to the bounding box containing it (δ_b).

An example of calculation of these quantities is shown in Fig. 10.9. Figure 10.9a and b show a representative picture a natural sample and the relative concentration (SiO_2) histogram, respectively. The maximum frequency of the histogram (i.e. the modal value) corresponds to $SiO_2 = 72.63$ wt.%. Figure 10.9c shows the concentration contours corresponding to this modal value; the bounding box ($\delta_b = 3.3 \cdot 4 = 13.2$ cm) used to normalize the length of contours (P_{tot}) is given by the perimeter of the image.

Considering the experimental data reported in Catrakis and Dimotakis (1998) the relation between $\log(Re)$ and $\log(P_{tot})$ is

$$\log\left(\frac{P_{tot}}{\delta_b}\right) = -0.33\log(Re) + 1.90 \quad (10.6)$$

from which $\log(Re)$ can be derived as

$$\log(Re) = \frac{1.90 - \log\left(\frac{P_{tot}}{\delta_b}\right)}{0.33} \quad (10.7)$$

Re values estimated for the natural structures are in the range from 500 to 7000. Note that the

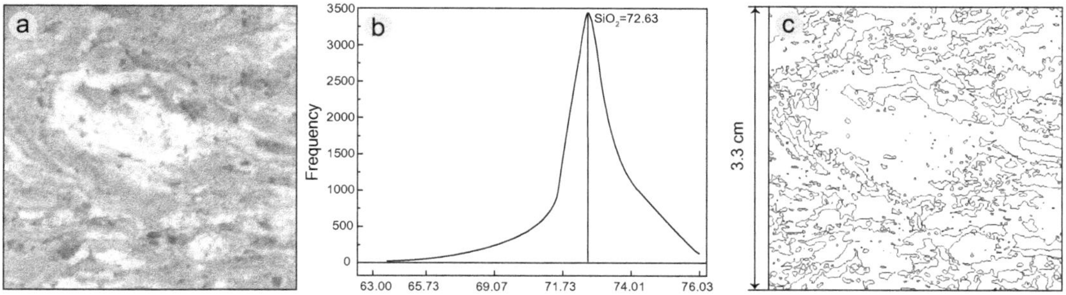

Fig. 10.9 **a** Representative natural sample; **b** concentration frequency histogram of the natural sample in **a**; **c** concentration contours traced from the image in **a** corresponding to the modal value of the histogram in **b**

length scale of the experiments is 10^{-2}-10^{-3} m, which agrees well with the length scale of the natural mixing patterns. Using the above equations to extrapolate larger ($Re > 10^4$) or lower ($Re < 10^3$) Reynolds numbers than those studied in Catrakis and Dimotakis (1998) requires caution because of the possible deviations from linearity of these empirical relationships.

Figure 10.10 shows that there is a direct relationship between the percentage of initial mafic magma estimated in natural samples and $\log(Re)$. We recall that the initial percentage of the mafic magma can be estimated using the two end-member mixing equation by knowing the initial concentration of chemical elements in the two end-members and the hybrid concentration, the latter estimated as the modal values of compositional histograms (Chap. 9). This is in keeping

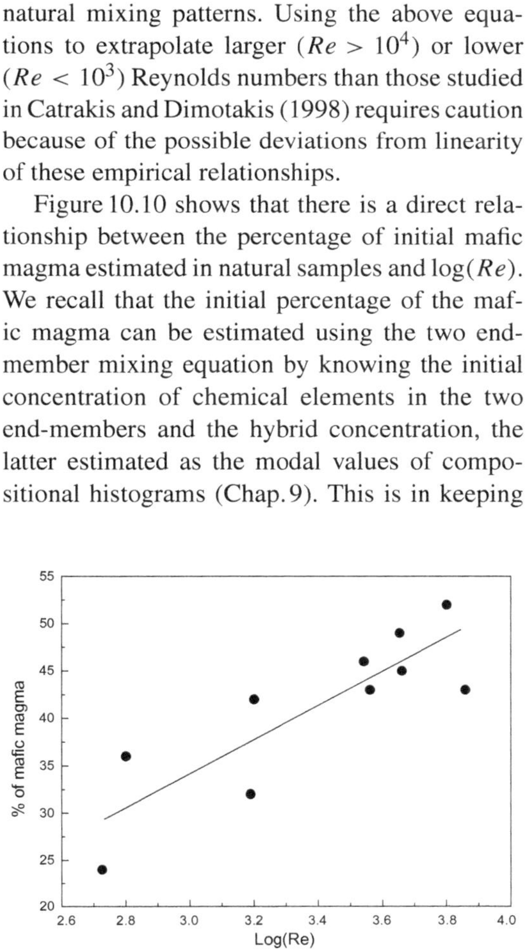

Fig. 10.10 Plot showing the correlation between the percentage of mafic magma in the natural samples and $\log(Re)$

with the fact that the higher is the percentage of the low viscosity mafic magma, the lower is the viscosity of the whole system. Turbulence is strongly favored in lower viscosity systems because Re is inversely correlated to viscosity as

$$Re_{\text{turb}} = \frac{\sqrt{q^2}L}{v} >> 1 \qquad (10.8)$$

where $\sqrt{q^2}$ is the characteristic velocity in the turbulent flow associated with the velocity fluctuations, Re_{turb} is the Reynolds number, L is the characteristic length of the system and μ is the kinematic viscosity (e.g. Raynal and Gence (1997)).

Results in Fig. 10.10 are in agreement with fluid dynamic experiments on magma mixing (e.g. Turner and Campbell 1986; Blake and Campbell 1986), showing that an increase of the volume of the mafic magma can trigger an increase in the turbulent regime of the mixing system.

Turbulent flows can be regarded as constituted by hierarchical cascades of eddies over a wide range of length scales. This can be described by an energy spectrum of velocity fluctuations at the different length scales.

As we have seen in Chap. 7, also the concentration distribution of the mixed patterns in the studied pumice samples can be described by a scale-invariant spectrum of concentrations (Fig. 7.21).

The analysis of concentration fields of mixed juvenile fragments indicated two main features: (i) the power spectrum can be described by a power-law over two orders of magnitude of wave numbers, corresponding to length scales from 0.1 to

10.0 cm (Fig. 7.21), and (ii) the frequency distribution of spectral slopes is normally distributed around the value ca. – 5/3 (Fig. 7.22).

The obtained results point to the scale-invariance of the scalar field (i.e. concentration) in the investigated range of length scales and are in general agreement with the concept of cascade in the theory of non-linear dynamics. This concept was developed by Kolmogorov et al. (1941) to describe the kinetic energy spectrum of turbulence. According to Kolmogorov et al. (1941), the turbulent spectrum of energy $E(k)$ can by described by the rate of transfer of energy from large to small scales. This flux correlates with the mean dissipation ε of kinetic energy into heat. In this context, the following relation holds

$$E(k) = C\varepsilon^{-2/3}k^{-5/3} \quad (10.9)$$

where C is a dimensionless constant. Obukhov (1949) and Corrsin et al. (1915) developed a similar theory for a passive scalar θ (e.g. concentration). In this case, the spectrum $F(k)$ of the passive scalar becomes

$$F(k) = C_\theta\varepsilon^{-1/3}\varepsilon_\theta k^{-5/3} \quad (10.10)$$

where k denotes the wavenumbers, C_θ is a constant, and ε and ε_θ are the average dissipation rates of the kinetic energy and the scalar variance, respectively. Therefore, following Eq. 10.10, for a tracer dispersed by turbulent flow there should be a spectrum $F(k) \propto k^{-5/3}$, which holds over a range of scales. This scaling relation is valid, however, only in the so-called "inertial-convective regime", which includes the range of scales larger than the viscous cut-off (i.e. those scales where the effect of viscosity and scalar diffusivity are both insignificant).

Comparing theoretical arguments and results from the analysis of concentration patterns in mixed rocks, we can highlight significant points of convergence. The first is that the spectral relation $F(k) \propto k^{-5/3}$ appears to hold also for natural samples: the concentration cascade (i.e. the spectrum of frequency distribution of concentration) is normally distributed around the modal value of −5/3 (Fig. 7.22). The second concerns the scaling range in concentration spectra; the slope of −5/3 is valid for natural data in the spectrum of length scales from 10 to 0.1 cm (Fig. 7.21). These length scales are generally larger than the length scale at which chemical diffusion is expected to operate in magmas (of the order of 10–100 μm). However, this is true only if we consider short time scales, as in the case of volcanic explosions rapidly triggered by magma mixing, where the mixing-to-eruption time must have been very short to prevent complete homogenization.

The issue of volcanic eruption time scales is a central point in volcanology and the crucial question is: how short is short? We will attempt to answer this question in the next chapter with the help of the compositional heterogeneity preserved in the rocks and the application of high-temperature experiments.

References

Bacon CR (1986) Magmatic inclusions in silicic and intermediate volcanic rocks. J Geophys Res 91(B6):6091. ISSN: 0148-0227. https://doi.org/10.1029/jb091ib06p06091

Beier C, Haase KM, Hansteen TH (2006) Magma Evolution of the Sete Cidades Volcano, São Miguel, Azores. J Petrol 47(7):1375–1411. ISSN: 1460-2415

Blake S, Campbell IH (1986) The dynamics of magma-mixing during flow in volcanic conduits. Contrib Miner Petrol 94(1):72–81

Catrakis HJ, Dimotakis PE (1998) Shape complexity in turbulence. Phys Rev Lett 80(5):968–971

Cioni R et al (1995) Compositional layering and syneruptive mixing of a periodically refilled shallow magma chamber: the AD 79 Plinian eruption of Vesuvius. J Petrol 36:739–750

Clocchiatti R et al (1994) Assessment of a shallow magmatic system: the 1888-90 eruption, Vulcano Island, Italy. Bull Volcanol 56(6-7):466–486. ISSN: 0258-8900. https://doi.org/10.1007/BF00302828

Clynne MA (1999) A complex magma mixing origin for rocks erupted in 1915, Lassen Peak, California. J Petrol 40(1):105–132. ISSN: 0022-3530. https://doi.org/10.1093/petroj/40.1.105

Corrsin S (1951) On the spectrum of isotropic temperature fluctuations in an isotropic turbulence. J Appl Phys 22(4):469–473. ISSN: 00218979. https://doi.org/10.1063/1.1699986

Davydova VO, Shcherbakov VD, Yu Plechov P (2018) The timescales of magma mixing in the plumbing system of Bezymianny Volcano (Kamchatka): insights from diffusion chronometry. Moscow Univ Geol Bull 73(5):444–450. ISSN: 19348436. https://doi.org/10.3103/S0145875218050058

De Astis G et al (2006) Geological map of the island of Vulcano (Aeolian Islands), scale 1:10,000. In: Litographia Artistica Cartographica, Florence, Italy

Druitt TH et al (2012) Decadal to monthly timescales of magma transfer and reservoir growth at a caldera volcano. Nature 482(7383):77-U97

Folch A, Marti J (1998) The generation of overpressure in felsic magma chambers by replenishment. Earth Planet Sci Lett 163(1–4):301–314. ISSN: 0012821X. https://doi.org/10.1016/S0012-821X(98)00196-4

Frazzetta G, La Volpe L (1991) Volcanic history and maximum expected eruption at "La Fossa di Vulcano" (Aeolian Islands, Italy). Acta Vulcanologicy 1:107–113

Gardner JE et al (1995) Influence of magma composition on the eruptive activity of Mount St Helens,Washington. Geology 23(6):523–530. ISSN: 00917613. https://doi.org/10.1130/0091-7613(1995)023

Gardner JE (2007) Heterogeneous bubble nucleation in highly viscous silicate melts during instantaneous decompression from high pressure. Chem Geol 236(1–2):1–12. ISSN: 00092541. https://doi.org/10.1016/j.chemgeo.2006.08.006

Gardner JE, Hélène Denis M (2004) Heterogeneous bubble nucleation on Fe-Ti oxide crystals in high-silica rhyolitic melts. Geochimica et Cosmochimica Acta 68(17):3587–3597. ISSN: 00167037. https://doi.org/10.1016/j.gca.2004.02.021

Gerbe MC et al (1992) Mineralogical and geochemical evolution of the 1982-1983 Galunggung eruption (Indonesia). Bull Volcanol 54(4):284–298. ISSN: 02588900. https://doi.org/10.1007/BF00301483

Gertisser R et al (2021) Processes and timescales of magma genesis and differentiation leading to the great tambora eruption in 1815. J Petrol 53(2):271–297. ISSN: 0022-3530. https://doi.org/10.1093/petrology/egr062

Giordano D, Russell JK, Dingwell DB (2008) Viscosity of magmatic liquids: a model. Earth Planet Sci Lett 271(1–4):123-134

Gualda GAR, Ghiorso MS (2007) Magnetite scavenging and the buoyancy of bubbles in magmas. Part 2: Energetics of crystal-bubble attachment in magmas. Contrib Mineral Petrol 154(4):479–490. ISSN: 00107999. https://doi.org/10.1007/s00410-007-0206-8

Hibbard MJ (1994) Petrography to petrogenesis. Prentice Hall

Hurwitz S, Navon O (1994) Bubble nucleation in rhyolitic melts: experiments at high pressure, temperature, and water content. Earth Planet Sci Lett 122(3–4):267–280. ISSN: 0012821X. https://doi.org/10.1016/0012-821X(94)90001-9

Kent AJR et al (2010) Preferential eruption of andesitic magmas through recharge filtering. English. In: Nature Geoscience 3(9):631–636

Kolmogorov A et al. (1941) The local structure of turbulence in incompressible viscous fluid for very Large Reynolds' numbers". DoSSR 30:301–305. ISSN: 0002-3264. https://ui.adsabs.harvard.edu/abs/1941DoSSR..30..301K/abstract

Kratzmann DJ et al (2009) Compositional variations and magma mixing in the 1991 eruptions of Hudson volcano, Chile. Bull Volcanol 71(4):419–439. ISSN: 14320819. https://doi.org/10.1007/s00445-008-0234-x

Leonard GS et al (2002) Basalt triggering of the c. AD 1305 Kaharoa rhyolite eruption, Tarawera Volcanic Complex, New Zealand. J Volcanol Geotherm Res 115(3–4):461–486. ISSN: 03770273. https://doi.org/10.1016/S0377-0273(01)00326-2

Mader HM, Llewellin EW, Mueller SP (2013) The rheology of two-phase magmas: a review and analysis. J Volcanol Geotherm Res 257:135–158. ISSN: 03770273. https://doi.org/10.1016/j.jvolgeores.2013.02.014

Mangan M, Sisson T (2000) Delayed, disequilibrium degassing in rhyolite magma: decompression experiments and implications for explosive volcanism. Earth Planet Sci Lett 183(3–4):441–455. ISSN: 0012821X. https://doi.org/10.1016/S0012-821X(00)00299-5

Marsh BD (1981) On the crystallinity, probability of occurrence, and rheology of lava and magma. Contrib Mineral Petrol 78(1):85–98. ISSN: 0010-7999. https://doi.org/10.1007/BF00371146

Martin VM et al (2008) Bang! Month-scale eruption triggering at Santorini volcano. Science 321(5893):1178

Murphy MD et al (1998) The role of magma mixing in triggering the current eruption at the Soufriere Hills Volcano, Montserrat,West Indies. Geophys Res Lett 25(18):3433–3436. ISSN: 00948276. https://doi.org/10.1029/98GL00713

Obukhov AM (1949) The structure of the temperature field in a turbulent flow. Izv. Akad. Nauk. SSSR, Ser. Geogr. Geophys. 13(58)

Oldenburg CM et al (1989) Dynamic mixing in magma bodies - theory, simulations, and implications. J Geophys Res-Solid Earth Planets 94(B7):9215–9236

Pallister JS, Hoblitt RP, Reyes AG (1992) A basalt trigger for the 1991 eruptions of Pinatubo volcano? Nature 356(6368):426-28. ISSN: 00280836. https://doi.org/10.1038/356426a0

Paredes-Mariño J et al (2017) Enhancement of eruption explosivity by heterogeneous bubble nucleation triggered by magma mingling. Sci Rep 7(1):1–10. ISSN: 20452322. https://doi.org/10.1038/s41598-017-17098-3

Perugini D, Poli G (2005) Viscous fingering during replenishment of felsic magma chambers by continuous inputs of mafic magmas: Field evidence and fluid-mechanics experiments. Geology 33(1):5–8

Perugini D et al (2015) Concentration variance decay during magma mixing: a volcanic chronometer. Sci Rep 5. ISSN: 20452322. https://doi.org/10.1038/srep14225

Piochi M et al (2009) Constraining the recent plumbing system of Vulcano (Aeolian Arc, Italy) by textural, petrological, and fractal analysis: the 1739 A.D. Pietre Cotte lava flow. Geochem Geophys Geosyst 10(1). ISSN: 15252027. https://doi.org/10.1029/2008GC002176

Queiroz G et al (2015) Eruptive history and evolution of sete cidades volcano, são miguel Island, Azores.

Geolog Soc Memoir. 44(1):87–104. Geological Society of London. https://doi.org/10.1144/M44.7
Raynal F, Gence JN (1997) Energy saving in chaotic laminar mixing. Int J Heat Mass Transf 40(14):3267–3273
Reagan MK (2003) Time-scales of differentiation from mafic parents to rhyolite in north american continental arcs. J Petrol 44(9):1703–1726. ISSN: 1460-2415. https://doi.org/10.1093/petrology/egg057
Robin C, Camus G, Gourgaud A (1991) Eruptive and magmatic cycles at Fuego de Colima volcano (Mexico). J Volcanol Geotherm Res 45(3–4):209–225. ISSN: 03770273. https://doi.org/10.1016/0377-0273(91)90060-D
Roscoe R (1952) The viscosity of suspensions of rigid spheres. Brit J Appl Phys 3(8):267–269. ISSN: 0508-3443. https://doi.org/10.1088/0508-3443/3/8/306
Self S, Rampino MR (2012) The 1963-1964 eruption of Agung volcano (Bali, Indonesia). Bull Volcanol 74(6):1521–1536. ISSN: 02588900. https://doi.org/10.1007/s00445-012-0615-z
Sigmundsson F et al (2010) Intrusion triggering of the 2010 Eyjafjallajökull explosive eruption. Nature 468(7322):426–432. ISSN: 00280836. https://doi.org/10.1038/nature09558
Smith PM, Asimow PD (2005) Adiabat_1ph: A new public front-end to the MELTS, pMELTS, and pHMELTS models. Geochem Geophys Geosyst 6(2):n/a–n/a. ISSN: 15252027. https://doi.org/10.1029/2004GC000816
Snyder D (2000) Thermal effects of the intrusion of basaltic magma into a more silicic magma chamber and implications for eruption triggering. Earth Planet Sci Lett 175(3–4):257–273
Sparks RSJ, Marshall LA (1986) Thermal and mechanical constraints on mixing between mafic and silicic magmas. J Volcanol Geotherm Res 29(1–4):99–124. ISSN: 03770273. https://doi.org/10.1016/0377-0273(86)90041-7
Sparks SRJ, Sigurdsson H, Wilson L (1977) Magma mixing: a mechanism for triggering acid explosive eruptions. Nature 267(5609):315–318. ISSN: 00280836. https://doi.org/10.1038/267315a0
Tepley FJ et al (2000) Magma mixing, recharge and eruption histories recorded in plagioclase phenocrysts from El Chichón Volcano, Mexico. J Petrol 41(9):1397–1411. ISSN: 1460-2415. https://doi.org/10.1093/petrology/41.9.1397
Turcotte DL (1992) Fractals and chaos in geology and geophysics. Cambridge University Press, Cambridge
Turner JS, Campbell IH (1986) Convection and mixing in magma chambers. Earth Sci Rev 23(4):255–352. https://doi.org/10.1016/0012-8252(86)90015-2
Vetere F et al (2013) Viscosity changes during crystallization of a shoshonitic magma: new insights on lava flow emplacement. J Mineral Petrol Sci 108(3):144–160. ISSN: 1345-6296. https://doi.org/10.2465/jmps.120724
N. Volynets O et al (1999) Holocene eruptive history of K-sudach volcanic massif, South Kamchatka: evolution of a large magmatic chamber. J Volcanol Geotherm Res 91(1–2):23–42. ISSN: 03770273. https://doi.org/10.1016/S0377-0273(99)00049-9
Vona A et al (2011) The rheology of crystal-bearing basaltic magmas from Stromboli and Etna. Geochimica et Cosmochimica Acta 75(11): 3214–3236. ISSN: 00167037. https://doi.org/10.1016/j.gca.2011.03.031
Williams SN, Self S (1983) The October 1902 plinian eruption of Santa Maria volcano, Guatemala. J Volcanol Geotherm Res 16(1–2):33–56. ISSN: 03770273. https://doi.org/10.1016/0377-0273(83)90083-5
Zanon V (2003) Magmatic feeding system and crustal magma accumulation beneath Vulcano Island (Italy): evidence from CO 2 fluid inclusions in quartz xenoliths. J Geophys Res 108(B6):2298. ISSN: 0148-0227. https://doi.org/10.1029/2002JB002140

A Geochemical Clock to Measure Timescales of Volcanic Eruptions

11

Not if... but when.

Austin Chambers, "Cascadia Fallen: Tahoma's Hammer"

Abstract

The eruption of volcanoes appears one of the most unpredictable phenomena on Earth. Quantification of the eruptive record constrains what is possible in a volcanic system. Timing is the hardest part to quantify. One of the main processes triggering explosive eruptions is the refilling of a sub-volcanic magma chamber by a new magma coming from depth. This process results in mixing and provokes a time-dependent diffusion of chemical elements. An eruption can initiate at any time during this process. Magma mixing is then frozen in time by the eruption. The time elapsed between mixing and eruption is recorded into the compositional patterns of the rocks. In this chapter we discuss a possible method that might help us cutting the Gordian knot of the presently intractable problem of volcanic eruption timing using a surgical approach integrating geochemical and experimental data on magma mixing. If these timescales can be linked with geophysical signals occurring prior to eruptions, the approach presented here may equip us with a new conceptual framework to move further steps towards one of the key objectives of volcanology: the prediction of volcanic eruptions.

11.1 Introduction

The eruption of volcanoes appears one of the most unpredictable phenomena on Earth. Yet the situation is rapidly changing. Quantification of the eruptive record constrains what is possible in a volcanic system. Timing is the hardest part to quantify.

As we have discussed in Chap. 10, we have the suspect that the main process triggering explosive eruptions is the refilling of a sub-volcanic magma chamber by a new magma coming from depth (Fig. 11.1). This process results in mixing and provokes a time-dependent diffusion of chemical elements. An eruption can initiate at any time during this process. Magma mixing is then frozen in time by the eruption. The time elapsed between mixing and eruption is recorded into the compositional patterns of the rocks.

Understanding the time elapsed from mixing to eruption is fundamental to discerning pre-eruptive behaviour of volcanoes in order to mitigate the huge impact of volcanic eruptions on society and the environment.

D. Perugini, *The Mixing of Magmas*, Advances in Volcanology,
https://doi.org/10.1007/978-3-030-81811-1_11

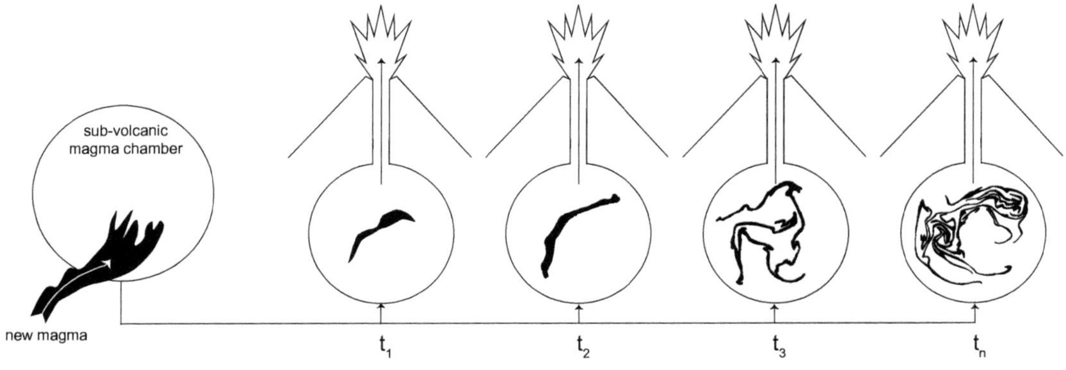

Fig. 11.1 Schematic representation of the replenishment of a magma chamber by a new magma and the subsequent mixing. An eruption can occur at any time (t) during mixing and the process will remain frozen in the erupted rocks

In this chapter we discuss a possible method that might help us cutting the Gordian knot of the presently intractable problem of volcanic eruption timing using a surgical approach integrating geochemical and experimental data on magma mixing.

Volcanic rocks are like a corpse. We cannot ask a corpse what was the cause of the decease. It would simply not answer. We can perform post-mortem examinations, an autopsy. And this is what we usually do as petrologists with rocks. We collect them, we take sections, we perform analyses. And, ultimately, we propose hypotheses. But we can do much more. We can wear the clothes of Dr. Frankenstein and resuscitate volcanic rocks in the laboratory with high-temperature experiments. Sometimes we create monsters, sometimes we discover new ways to look at the complexity of Nature. In any case, we can light a new candle to shed further light on another segment of road towards a better knowledge the way volcanoes work. We must try to light up as many new candles as we can to avoid getting stuck to conventional views and conceptual models. New ideas will allow us to get rid of the dangerous so-called "streetlight effect", which I wish to report here: a policeman sees a drunk man searching for something under a streetlight and asks what the drunk has lost. He says he lost his keys and they both look under the streetlight together. After a few minutes the policeman asks if he is sure he lost them here, and the drunk replies, no, and that he lost them in the park. The policeman asks why he is searching here, and the drunk replies, "this is where the light is" (Kaplan 1964).

The argument presented in this chapter may aid in shedding new light and open a new window on the physical-chemical processes occurring in the days preceding volcanic eruptions providing information to estimate eruption timescales. If these timescales can be linked with geophysical signals occurring prior to eruptions, the approach presented here may equip us with a new conceptual framework to move further steps towards one of the key objectives of volcanology: the prediction of volcanic eruptions.

11.2 Mixing-to-Eruption Timescales for Phlegrean Fields Volcanoes

The first case study that we consider is represented by eruptions belonging to the recent activity of the Phlegrean Fields caldera (Italy). Here the knowledge of eruption timescales is crucial since the area is home to more than 1.5 million people (Orsi et al. 2004). The recent volcanic activity of Phlegrean fields (60 ka BP to 1538 A.D.) was characterized by many explosive events (about seventy eruptions have been documented in the last 15 ka (e.g. Orsi et al. 2004). Phlegrean Fields are an active volcanic system that is likely to erupt in the future. Understanding mechanics and timing of eruptions is, therefore, a crucial issue.

Magma mixing processes have been widely documented in this volcanic system (e.g. Tonarini

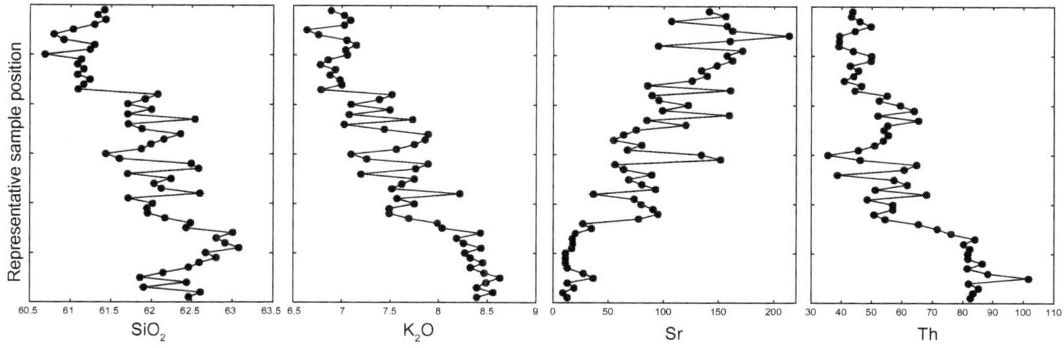

Fig. 11.2 Variation of some representative major and trace elements across the pyroclastic sequence of Averno; height of the sequence is ca. 20 m

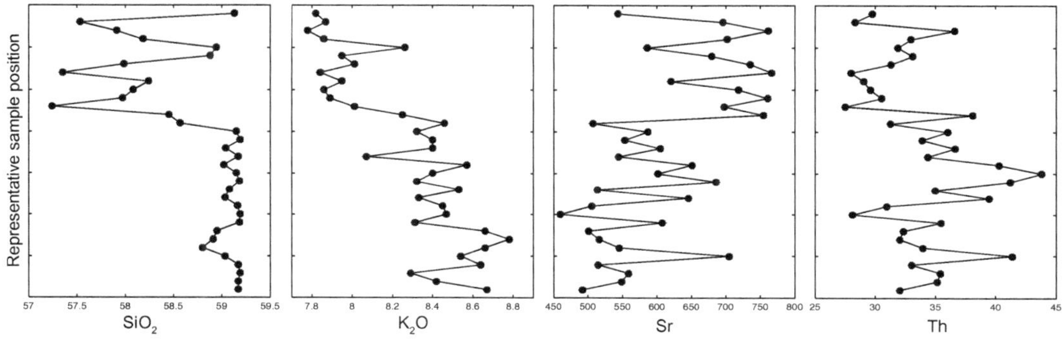

Fig. 11.3 Variation of some representative major and trace elements across the pyroclastic sequence of Astroni; height of the sequence is ca. 16 m

et al. 2009; Pappalardo et al. 2002; Astbury et al. 2018; Perugini et al. 2010). Here we concentrate on three eruptive sequences, erupted by distinct eruptive centres, known in the literature as Astroni 6, Averno 2, and Agnano-Monte Spina, generated by explosive activity that occurred in the III epoch (4.8–3.8 ka) of the Phlegrean Fields Caldera (e.g. Rosi et al. 1983; Di Vito et al. 1999; Isaia et al. 2004).

Figures 11.2, 11.3 and 11.4 display the variation of representative major and trace elements along the three stratigraphic sequences. They all show a clear chemical zonation. In the plots is displayed the actual relative position of the analysed samples. The chemical zoning in the magma chambers that were discharged by the eruptions can be approximately inferred reversing the stratigraphic sequence. Therefore, it can be said that the three systems were constituted by the presence of less evolved magmas at the bottom, overlaid by more evolved melts. The compositional zoning shows zig-zag patterns forming, in general, a monotonous variation towards the most evolved compositions and indicating interaction between the different compositions.

We described high-temperature magma mixing experiments performed with the centrifuge apparatus in Chap. 5. With this experimental device we simulated the injection and mixing of a less evolved melt into a more evolved one. The initial compositions of end-members were an alkali basalt and a phonolite, considered as the possible end-members that participated to the mixing process in the Phlegrean Fields volcanic systems (Civetta et al. 1997; Perugini et al. 2010, 2015). Results from these experiments can be used in the attempt of estimating the mixing-to-eruption time for the three case studies.

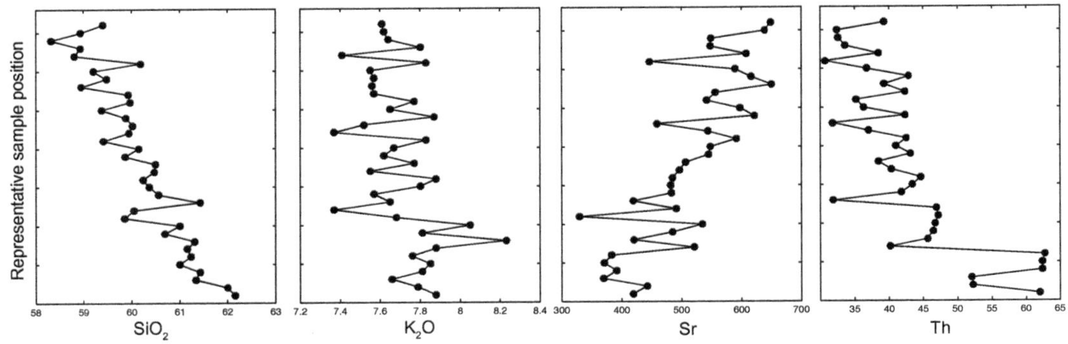

Fig. 11.4 Variation of some representative major and trace elements across the pyroclastic sequence of Agnano-Monte Spina; height of the sequence is ca. 15 m

Figure 11.5 shows the time evolution of the mixing process along compositional transects measured on the experimental samples (Fig. 5.2). As the mixing time increases, the compositional variation is progressively smoothed by the combined development of advection (stretching and folding) and diffusion.

As we discussed in Chaps. 4 and 5, concentration variance (σ_n^2) can quantify the time evolution of chemical exchanges during the course of mixing. Figure 11.6 shows the exponential decay of concentration variance for the three mixing experiments for some representative chemical elements. We can observe that σ_n^2 decreases rapidly, relaxing towards zero within 120 min. The concentration variance decay for all chemical elements was fitted using Eq. 4.6.

These empirical relationships were applied for all analysed elements of the three compositional sequences of Averno, Astroni and Agnano-Monte Spina (Figs. 11.2, 11.3 and 11.4) in order to estimate the timescale of magma mixing preceding eruptions (i.e. the mixing-to-eruption time interval).

Here we rely upon the assumption that we can use experimental data to make inferences about natural processes considering that the evolution in space and time of magma mixing is governed by chaotic dynamics, as extensively discussed in the previous chapters. We recall that stretching and folding processes, i.e. the basic forces promoting magma mixing, generate scale-invariant structures (i.e. fractal compositional patterns). This implies that the compositional patterns are repeated at many length scales (from the meter to the micrometre). This also means that, in principle, the analysis of magmatic systems at any length scale will provide, statistically, the same type of information. This conceptual model allows us to justify the use of small-scale experimental samples to investigate natural processes on a much larger range of scales.

Using the exponential empirical relationships obtained from the magma mixing experiments, the mixing time for each element was estimated. Results show that the mixing-to-eruption time was of the order of 18 $\pm$ 5 (s.d.), 13 $\pm$ 4 (s.d.) and 15 $\pm$ 4 (s.d.) minutes for Averno, Astroni and Agnano-Monte Spina eruptions, respectively. It is important to note that a large of number of chemical elements (of the order of thirty) were used to estimate these timescales. Nevertheless, they can be estimated with a large level of confidence, indicating the robustness of the method.

A further indication of the robustness of the approach relies on the fact that removing data randomly from the natural stratigraphic sequences (up to 75 % of samples) the time estimates remain within $\pm$4–5 min s.d (Perugini et al. 2015). As an example, for the Averno sequence, 15 samples collected randomly along the pyroclastic deposit are enough to apply the method illustrated above to estimate the mixing-to-eruption timescale.

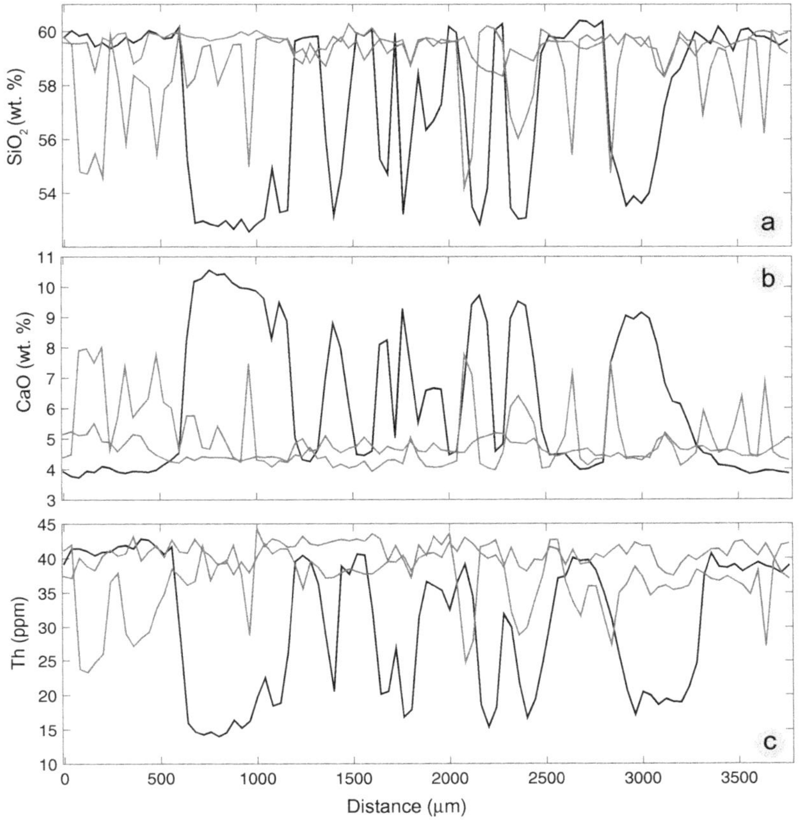

Fig. 11.5 Compositional variation across the mixing interface of some representative elements for the mixing experiments performed using the centrifuge apparatus (see Chap. 5). The different colors indicate experiments at different mixing times (black: 5 min, green: 20 min, red: 120 min)

The estimated timescales point to a very short time separating the beginning of mixing and eruption. Establishing whether magma mixing was the trigger or the consequence of the eruption is not easy. In any case, however, having the opportunity to measure these timescales represents an advancement in our knowledge about the time elapsing between the onset of chemical and physical instability in the sub-volcanic magma chamber and the eruption.

Phlegrean Fields shallow reservoirs have been suggested to be located at depths of four to five kilometers (e.g. Arienzo et al. 2010). Considering the estimated timescales, this implies average ascent velocities of the magmas of the order of 5–8 m/s. This indicates rapid ascent of magmas through the plumbing system. These ascent velocities are similar to those that are needed for sustained eruptive flows (Richard et al. 2013). Large ascent velocities are also in agreement with the fact that Phlegrean Fields magma bodies experienced magma mixing. The injection of hot and low viscosity mafic melts may have contributed to fluidize the whole magmatic mass (Kent et al.

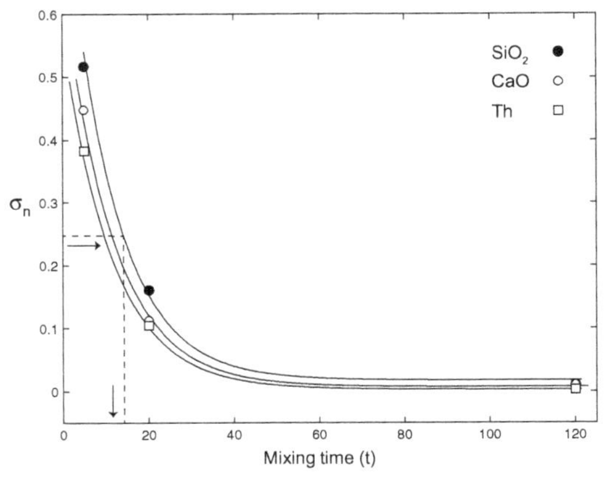

Fig. 11.6 Concentration variance decay for some representative elements fitted using Eq. 4.6

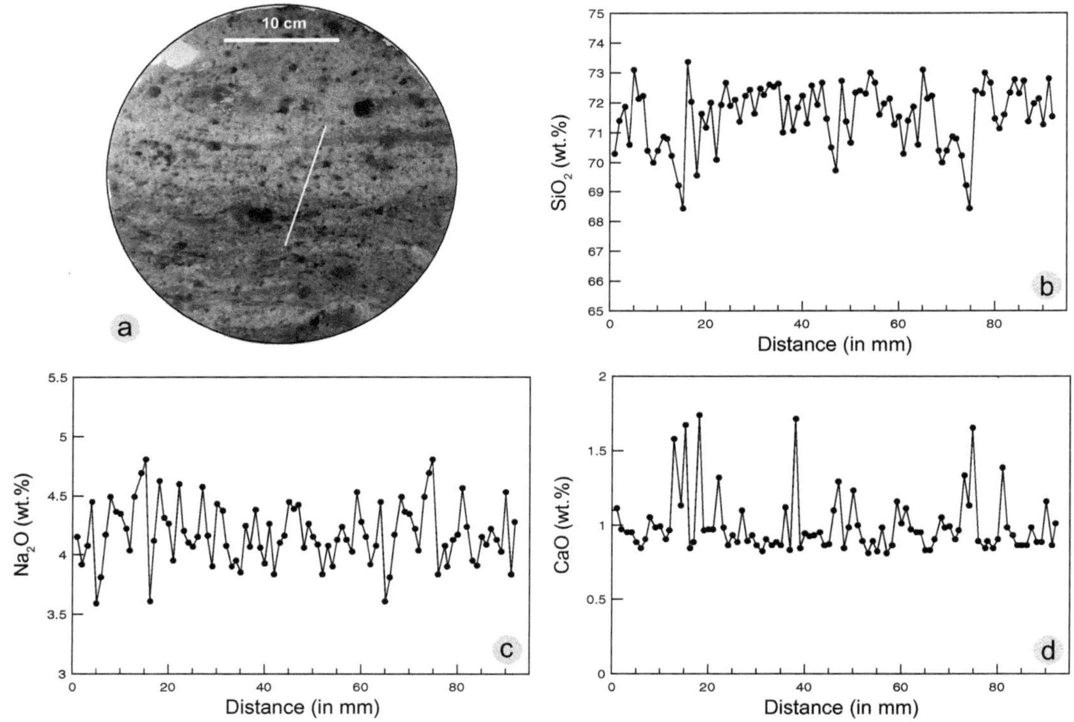

Fig. 11.7 **a** Representative natural sample from the Island of Vulcano showing compositional banding; **b–d** variation of the concentration of representative elements across the transect reported in **a** (white line)

2010) accelerating magma migration towards the Earth's surface.

11.3 Mixing-to-Eruption Timescales for the Island of Vulcano, Aeolian Archipelago

The second case is represented by the eruptive cycle of A.D. 1739 on the Island of Vulcano (Aeolian Archipelago, Italy), which we have already encountered in the study of magmatic enclaves occurring in the Pietre Cotte lava flow (Chap. 7).

The A.D. 1739 eruption developed in two main phases. The first phase was explosive and produced pyroclastic density currents and fallout deposits (Frazzetta and La Volpe 1991). These are prevalently exposed along the flanks of the volcanic edifice and contain banded pumices with trachytic and rhyolitic compositions. The second phase was effusive and erupted the Pietre Cotte lava flow.

Banded pumice samples were analysed for their major element contents along transects crossing the compositional banding. A representative sample is shown in Fig. 11.7. The compositional transects display alternate "up and down" patterns, as it should be expected on samples generated by intimate mixing of magmas.

Magma mixing experiments were performed using the COMMA device (Morgavi et al. 2015; see Chap. 5). Two natural compositions from the Island of Vulcano were used as end-members in the experiments. The least evolved is a latite cropping out in the upper part of La Fossa cone. The most evolved end-member is a high-K rhyolite from the Pietre Cotte lava flow (Rossi et al. 2019). The choice of using these two end-members was motivated by the assumption that they have been suggested as the most extreme compositions participating to the mixing process during A.D. 1739 eruption (Piochi et al. 2009; Vetere et al. 2015). Two experiments with different duration (10.5 h

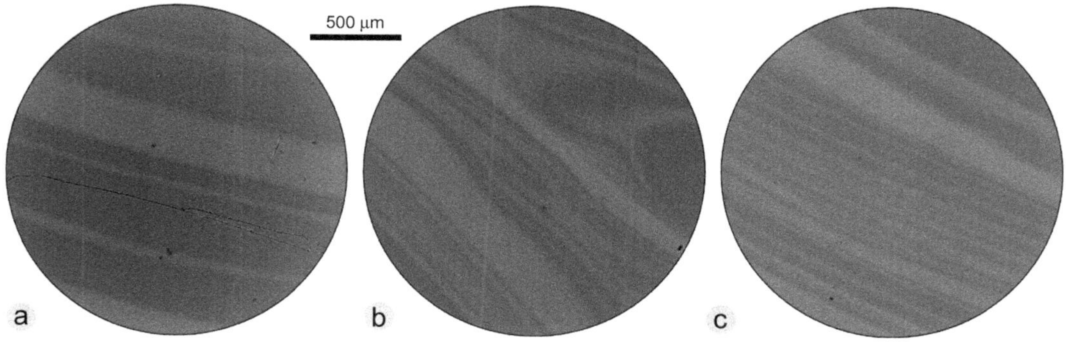

Fig. 11.8 BSE images of samples produced during experimental magma mixing using the chaotic magma mixing apparatus

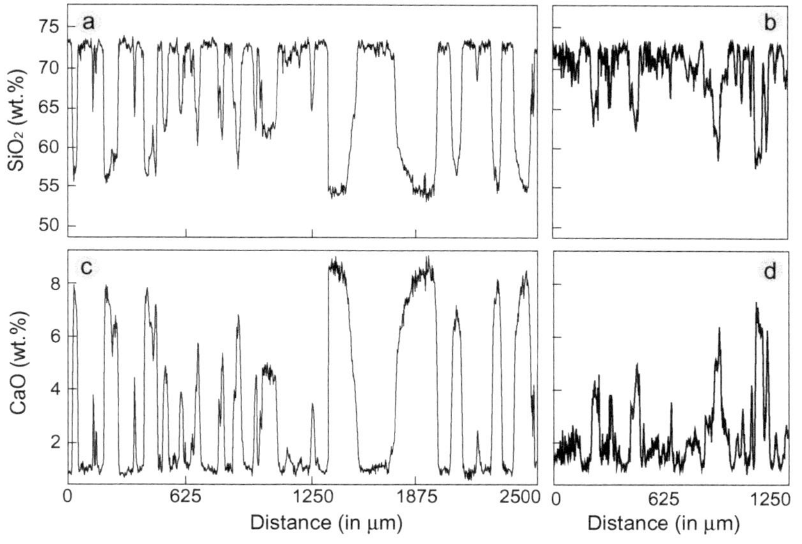

Fig. 11.9 Compositional variation of representative elements across the mixing patterns of the experiments performed using the COMMA; **a–c** experimental mixing time: 10.5 h; **b–d** experimental mixing time: 31.5 h

and 31.5 h) were performed. The applied experimental mixing protocol is reported in Morgavi et al. (2015) and Rossi et al. (2017).

Back-scattered electron images of representative portions of the experimental products are shown in Fig. 11.8. The samples show an intricate pattern of filaments with alternating compositions of the two melts.

The compositional variation of major elements across the experimental samples was measured along transects crossing the filaments. Representative transects for selected major elements are displayed in Fig. 11.9. The plots show, as expected, that the compositional fluctuations across the experimental samples decrease as mixing time increases, an effect of the progress of stretching and folding dynamics and chemical diffusion.

The concentration variance (σ_n^2) was estimated for the experimental products and the graph of Fig. 11.10 shows its variation as a function of time. As for the compositional data measured on the centrifuge experiments, σ_n^2 decays exponentially with increasing mixing time, with the different elements showing different rates of variation of σ_n^2. This is symptomatic of the occurrence of chaotic dynamics during magma mixing (see Chaps. 4 and 5).

Analogous to the case study of Phlegrean Fields, concentration variance was estimated for the compositional variability measured on the

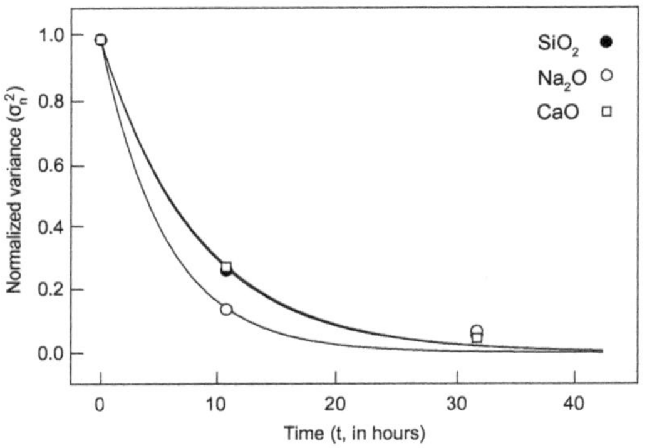

Fig. 11.10 Concentration variance decay for some representative elements fitted using Eq. 4.6

natural samples (Rossi et al. 2019). Comparison between variance values estimated on natural samples and the empirical relationships obtained experimentally allows us to estimate the mixing-to-eruption time for the A.D. 1739 eruption of La Fossa cone. Considering the uncertainty of the fitting of experimental data and the measurements on the natural sample, the estimated mixing-to-eruption time ranges from ca. 10 to 39 h, with an average of 29 $\pm$ 9 (s.d.) hours (Rossi et al. 2019).

As in the case of Phlegrean Fields, these timescales can be used to estimate the average velocity of magma ascent during the A.D. 1739 La Fossa cone eruption. De Astis et al. (2013) suggested, on the basis of petrological data, that possible storage regions of the magmas below the volcano can be situated at ca. 3–5 km depth. We assume that this is the region of the volcanic edifice were the injection of the trachytic magma into the rhyolitic one has occurred. This implies average magma ascent rates ranging from $3 \cdot 10^{-2}$ to $5 \cdot 10^{-2}$ ms^{-1}. This ascent velocities are significantly lower than those estimated for the eruption of the Phlegrean Fields in the previous section. There are many factors that can influence magma ascent velocity in different volcanic system and their discussion is outside the scope of this book. Nevertheless, we can hypothesize that one of the causes might reside in differences in the overall rheological behaviour of the two systems. The larger ascent velocity values for Phlegrean Fields can possibly be explained by the fact that K-alkaline magmas have lower viscosities compared to calc-alkaline magmas, as in the case of the Island of Vulcano. Noteworthy is the fact that ascent velocity estimated for the case of Vulcano are comparable to those estimated for calc-alkaline magmas (Castro and Dingwell 2009), indicating that the rheological behaviour of magmas can provide an approximate explanation.

11.4 Mixing-to-Eruption Timescales for Sete Cidades Caldera, Azores

The third and last case study we consider is represented by the Santa Barbara formation (Sao Miguel Island, Azores; ca. 16 ka) corresponding to the last eruptive phase of caldera formation of the Sete Cidades Volcano (Queiroz et al. 2015; Beier et al. 2006). We already considered this outcrop in the previous chapter when we studied the possible role of heterogeneous nucleation of bubbles on the mafic fragments. The Santa Barbara formation is one of the major explosive events that occurred in the Sete Cidades volcano. As discussed in the previous chapter, eruptive products are characterized by the presence of mafic enclaves dispersed within trachytic pumice clasts testifying for the occurrence of magma mixing processes (Queiroz et al. 2015; Paredes-Mariño et al. 2017; Laeger et al. 2019). That magma mixing occurred is also suggested by the fact that closer observation of some trachytic clasts reveals the presence of filaments with different colours corresponding to different compositions (Laeger et al. 2019).

This feature is shown in Fig. 11.11 where representative binary inter-elemental plots of EMPA analyses performed on these glasses are reported as a function of SiO_2. It is interesting to note how some elements, such as CaO tend to define ca. linear trends between the two hypothesised end-members (see below), whereas other elements, such as Na_2O, are completely scattered. We have already encountered this behaviour in Chap. 5 and we have shown that this is most probably due to

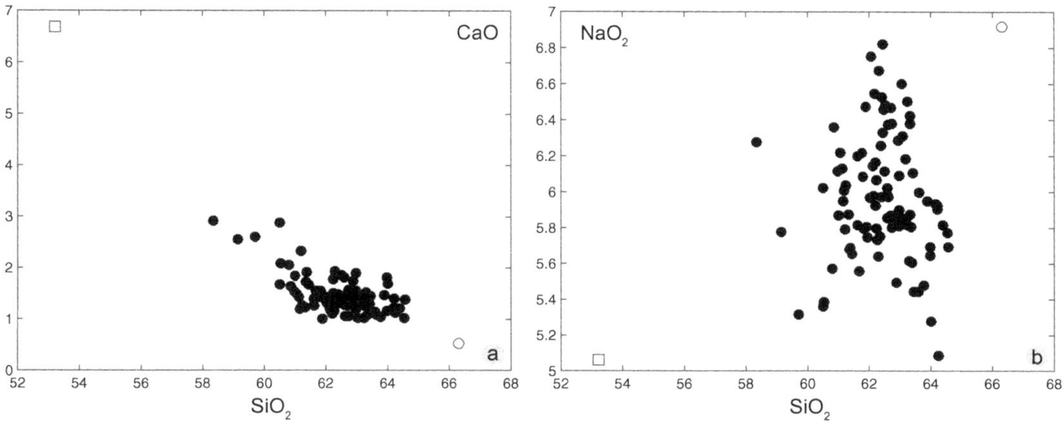

Fig. 11.11 Inter-elemental plots of representative chemical elements measured on the glasses of Santa Barbara formation pumices. Open symbols represent the mixing end-members

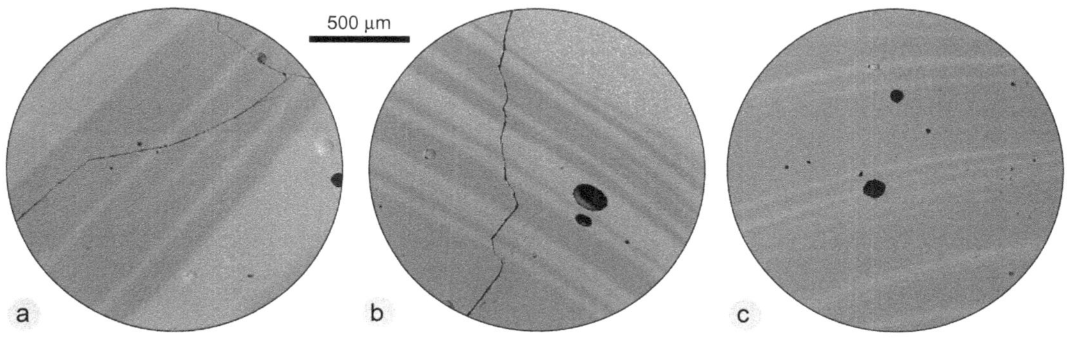

Fig. 11.12 BSE images of samples produced during experimental magma mixing using the chaotic magma mixing apparatus. Mixing time increases from **a** to **c**

the different mobility of chemical elements in the magmatic system during mixing.

As in the case of the Island of Vulcano, magma mixing experiments were performed using the COMMA device. In the experiments, two natural compositions, considered as plausible end-members for the mixing process at Sete Cidades volcano, were used. White trachytic pumices showing no evidence of mixing from the Santa Barbara outcrop were used as the felsic end-member. The mafic end-member was a mugearite cropping out as a dike on the caldera walls of Sete Cidades volcano (Laeger et al. 2019). Three experiments with different duration (ca. 10.5, 21 and 42 h) were performed using the same experimental mixing protocol reported in Morgavi et al. (2015) and Rossi et al. (2017).

Back-scattered electron images of representative portions of the experimental products are displayed in Fig. 11.12. As the mixing process progressed in time the two melts were intimately mixed producing an increasing number of filaments with different compositions in which chemical exchanges become progressively more efficient.

The compositional variation of major elements on the experimental samples was analysed along transects crossing the filaments; representative transects are displayed in Fig. 11.13 for the three experiments. Here we can see, as in the previous two case studies, that the compositional variability decreases as a function of mixing time.

The concentration variance (σ_n^2) was estimated for the experimental products. Figure 11.14 shows

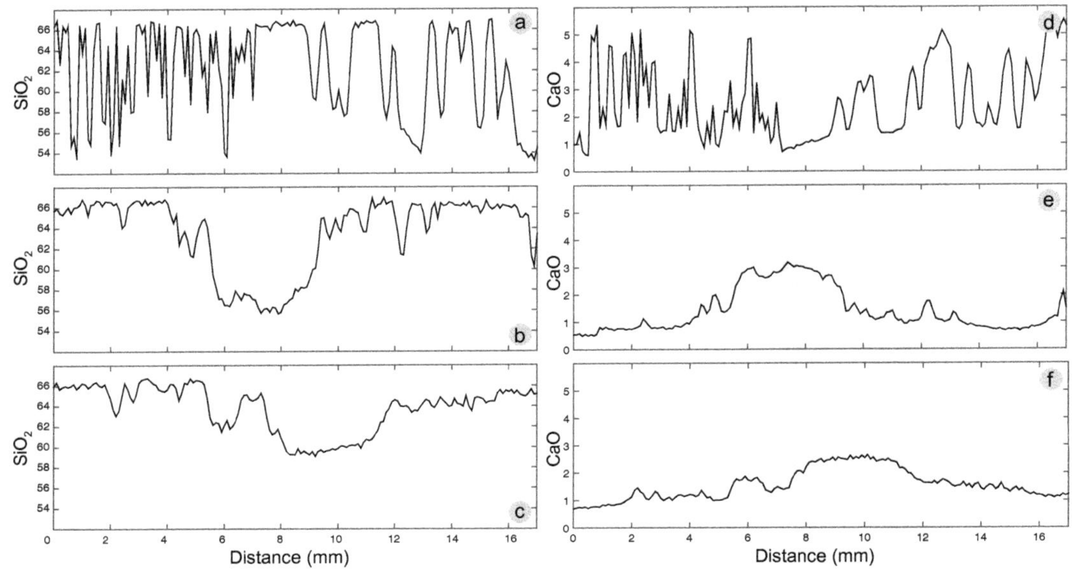

Fig. 11.13 Compositional variation of representative elements across the mixing patterns for the experiments performed using the COMMA; **a–d** experimental mixing time: 10.5 h; **b–e** experimental mixing time: 21 h; **c–f** experimental mixing time: 42 h

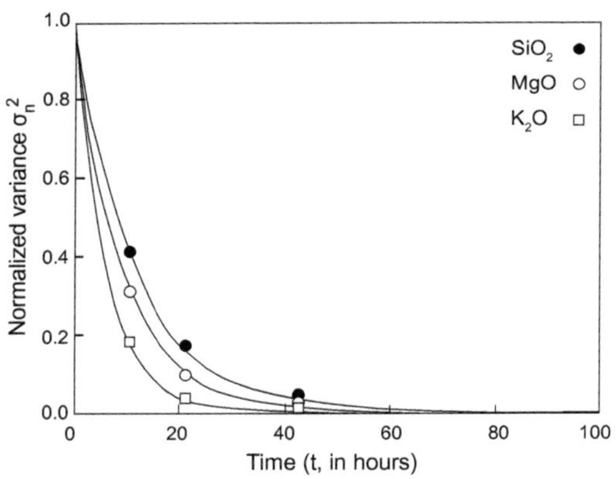

Fig. 11.14 Concentration variance decay for some representative elements fitted using Eq. 4.6

the variation of σ_n^2 in time. As in the previous cases, σ_n^2 decays exponentially with increasing mixing time, with the different elements showing different rates of variation of σ_n^2.

Concentration variance was estimated for the compositional variability measured on the natural samples (Laeger et al. 2019). Comparison between these values and the empirical relationships obtained experimentally allows us to estimate the mixing-to-eruption time for Santa Barbara eruption. This ranges between 19.0 and 40.6 h, with an average of 29 ± 8 h (Laeger et al. 2019). It is interesting to note that these timescales are of the same order of magnitude of those estimated for the A.D. 1739 eruption on the Island of Vulcano. Beier et al. (2006) suggested that the shallow reservoirs of the Sete Cidades volcano in which the mixing process occurred are possibly located at a depth of 3–5 km. This implies that average ascent rates for the magmas that generated the Santa Barbara explosions are similar to those of the A.D. 1738 eruption on the Island of Vulcano. The rheological behaviour of Santa Barbara magmas in not very dissimilar from that of the magmas which erupted during the Pietre Cotte eruption. This supports the general idea that magma rheology might provide an explanation for the different ascent rates between Phlegrean Fields eruptions and those occurred at Sao Miguel and Vulcano.

In this last part of the book, we put forward the idea that the evidence of magma mixing in the rocks can be exploited to obtain information about the timing of volcanic eruptions. As we have discussed at the beginning of the previous chapter, magma mixing appears to be a common factor of most explosive eruptions on Earth. It would appear logical, therefore, to infer that the methods reported here can be easily applied to derive the temporal information for most volcanoes. If, in principle, this might appear feasible, we need to control our enthusiasm and consider the limits of this approach. Among them, probably the most important are: (i) experiments were performed with natural silicate melts at superliquidus conditions; although this limit was mitigated in the studied natural cases because of the low crystal content, we must be aware that most magmas may contain larger amount of crystals, whose effect on the rheological behaviour of magmas is not easy to take into account in the experiments and the analysis of natural samples; (ii) the identification of the end-members that participated to the mixing process is often not an easy task. Detailed petrological and geochemical investigations are needed in order to reduce as much as possible the uncertainty in their identification; (iii) the geological record might not be complete, introducing additional sources of uncertainty in both recovering erupted samples and reconstruction of eruptive dynamics.

Therefore, additional conceptual and technical efforts must be made in order to build a solid methodological approach that can be utilized as a standard practice for the determination of eruptive timescales. While the identification of the end-members and the recovering of erupted samples can be sometimes a challenging, yet feasible task, much larger efforts must be done to design a new generation of mixing experiments including crystals in the end-member magmas. This is not trivial since the introduction of crystals will strongly amplify the complexity of the system, requiring, for example, larger experimental setups in order not to incur in side-wall effects that will be certainly magnified due to the presence of solid phases.

Moving further steps, therefore, will require time, highly skilled researchers, additional funding and, hopefully, a world-wide joint effort to put together the best petrologists and volcanologists working towards a common goal.

We wish to imagine that in the future, the method proposed here in its embryonic form, may provide the basis for developing a new tool to be used to build an historical inventory of eruption timescales for planet Earth, providing information of vital importance for the mitigation of associated risks. This historical inventory may have an immense value if eruption timescales can be linked with geophysical signals (i.e. seismicity, gravity changes and ground deformations) preceding eruptions in active volcanoes because we may be able to temporally track the development of explosive eruptions starting from magma chamber dynamics and possibly mitigate their impact on environment and society.

References

Arienzo I et al (2010) The feeding system of Agnano-Monte Spina eruption (Campi Flegrei, Italy): dragging the past into present activity and future scenarios. Chem Geol 270(1–4):135–147

Astbury RL et al (2018) Tracking plumbing system dynamics at the Campi Flegrei caldera. Highresolution trace element mapping of the Astroni crystal cargo, Italy

Beier C, Haase KM, Hansteen TH (2006) Magma evolution of the Sete Cidades volcano, São Miguel, Azores. J Petrol 47(7):1375–1411. issn: 1460-2415. https://doi.org/10.1093/petrology/egl014

Castro JM, Dingwell DB (2009) Rapid ascent of rhyolitic magma at Chaiten volcano, Chile. Nature 461(7265):780–U29

Civetta L et al (1997) Geochemical zoning, mingling, eruptive dynamics and depositional processes - the Campanian Ignimbrite, Campi Flegrei caldera, Italy. J Volcanol Geotherm Res 75(3–4):183–219

De Astis G et al (2013) Geology, volcanic history and petrology of vulcano (central aeolian archipelago). Geol Soc Mem 37(1):281–349. issn: 04354052. https://doi.org/10.1144/M37.11

Di Vito MA et al (1999) Volcanism and deformation since 12,000 years at the Campi Flegrei caldera (Italy). J Volcanol Geotherm Res 91(2–4):221–246

Frazzetta G, La Volpe L (1991) Volcanic history and maximum expected eruption at "La Fossa di Vulcano" (Aeolian Islands, Italy). Acta Vulcanol 1:107–113

Isaia R et al (2004) The Astroni volcano: the only example of closely spaced eruptions in the same vent area during the recent history of the Campi Flegrei caldera (Italy). J Volcanol Geotherm Res 133(1–4):171–192. issn: 03770273. https://doi.org/10.1016/S0377-0273(03)00397-4

Kaplan A (1964) The conduct of inquiry: methodology for behavioral science. Transaction Publishers

Kent AJR et al (2010) Preferential eruption of andesitic magmas through recharge filtering. Nat Geosci 3(9):631–636

Laeger K et al (2019) Pre-eruptive conditions and triggering mechanism of the 16 ka Santa Barbara explosive eruption of Sete Cidades Volcano (Sao Miguel, Azores). Contrib Mineral Petrol 174(2). issn: 00107999. https://doi.org/10.1007/s00410-019-1545-y

Morgavi D et al (2015) High-temperature apparatus for chaotic mixing of natural silicate melts. https://doi.org/10.1063/1.4932610

Orsi G, Antonio Di Vito M, Isaia R (2004) Volcanic hazard assessment at the restless Campi Flegrei caldera. Bull Volcanol 66(6):514–530. issn: 02588900. https://doi.org/10.1007/s00445-003-0336-4

Pappalardo L et al (2002) Evidence for multi-stage magmatic evolution during the past 60 kyr at Campi Flegrei (Italy) deduced from Sr, Nd and Pb isotope data. J Petrol 43(8):1415–1434

Paredes-Mariño J et al (2017) Enhancement of eruption explosivity by heterogeneous bubble nucleation triggered by magma mingling. Sci Reports 7(1):1–10. issn: 20452322. https://doi.org/10.1038/s41598-017-17098-3

Perugini D et al (2015) Concentration variance decay during magma mixing: a volcanic chronometer. Sci Rep 5. issn: 20452322. https://doi.org/10.1038/srep14225

Perugini D et al (2010) Time-scales of recent Phlegrean fields eruptions inferred from the application of a 'diffusive fractionation' model of trace elements. Bull Volcanol 72(4):431–447. https://doi.org/10.1007/s00445-009-0329-z

Piochi M et al (2009) Constraining the recent plumbing system of Vulcano (Aeolian Arc, Italy) by textural, petrological, and fractal analysis: the 1739 A.D. Pietre Cotte lava flow. Geochem Geophys Geosyst 10(1). issn: 15252027. https://doi.org/10.1029/2008GC002176

Queiroz G et al (2015) Eruptive history and evolution of sete cidades volcano, são miguel Island, Azores. Geolog Soc Mem 44(1):87–104. Geological Society of London. https://doi.org/10.1144/M44.7

Richard D et al (2013) Outgassing: influence on speed of magma fragmentation. J Geophys Res Solid Earth 118(3):862–877. issn: 21699313. https://doi.org/10.1002/jgrb.50080

Rosi M, Sbrana A, Principe C (1983) The phlegraean fields: structural evolution, volcanic history and eruptive mechanisms. J Volcanol Geotherm Res 17(1–4):273–288. issn: 03770273. https://doi.org/10.1016/0377-0273(83)90072-0

Rossi S et al (2017) Exponential decay of concentration variance during magma mixing: robustness of a volcanic chronometer and implications for the homogenization of chemical heterogeneities in magmatic systems. Lithos 286–287. issn: 18726143. https://doi.org/10.1016/j.lithos.2017.06.022

Rossi S et al (2019) Role of magma mixing in the pre-eruptive dynamics of the Aeolian Islands volcanoes (Southern Tyrrhenian Sea, Italy). Lithos 324–325. issn: 18726143. https://doi.org/10.1016/j.lithos.2018.11.004

Tonarini S et al (2009) Geochemical and B-Sr-Nd isotopic evidence for mingling and mixing processes in the magmatic system that fed the Astroni volcano (4.1-3.8 ka) within the Campi Flegrei caldera (southern Italy). Lithos 107(3–4):135–151. issn: 00244937. https://doi.org/10.1016/j.lithos.2008.09.012

Vetere F et al (2015) Dynamics and time evolution of a shallow plumbing system: the 1739 and 1888-90 eruptions, Vulcano Island, Italy. J Volcanol Geotherm Res 306. issn: 03770273. https://doi.org/10.1016/j.jvolgeores.2015.09.024

Concluding Remarks

Yet it is no easy task.
The classification of the constituents of a chaos,
nothing less is here essayed.

Hermann Melville, "Moby Dick"

The aim of this book was to show the multifaceted nature of magma mixing using, together with classical petrological techniques, also the non-conventional methods of Chaos Theory and Fractal Geometry.

The combined use of these techniques allowed us to draw a picture of the complex chemical and physical mechanisms that characterize mixing processes from the microscale to the macroscale. In particular, it was highlighted how the presence of chaotic dynamics and the consequent generation of fractal domains are ubiquitous in magma mixing systems at many length scales.

The study of magma mixing in volcanic rocks provided the opportunity to observe snapshots of the evolution of this process in time and space. In particular, we have been able to recognize two main types of dynamical regions coexisting at many length scales within the same system: i) chaotic regions, characterized by a very efficient dispersion of magmas and ii) regular regions in which magmas are little deformed. In these two regions chemical exchanges between magmas develop with different styles and efficiencies. In chaotic regions the interfacial area between magmas increases exponentially, generating compositional gradients at progressively finer scales in a short time and allowing chemical exchanges to be very efficient, producing large degrees of hybridization. In regular regions the interfacial area between the magmas remains approximately constant and chemical diffusion is generally inefficient, allowing these portions of magmas to survive complete hybridization. These results allowed us to hypothesize that magmatic systems can show large degrees of spatial and temporal chemical inhomogeneity at many observational scales. This is essentially due to the kinematics of the flow fields developing within the magmatic masses and their fractal structure.

Magma mixing patterns were analyzed using the methodological approaches made available from Chaos Theory and Fractal Geometry. Results showed that not only magma mixing can be regarded as a chaotic process, but that we can quantify the degree of "chaoticity" of mixing patterns using fractal geometry techniques.

Starting from natural rocks, the mixing process was simulated numerically and with the help of high-temperature experiments performed with new equipment that were specifically designed to address the complexity of magma mixing. The use of these systems allowed us to reproduce, as a first approximation, the patterns observed in natural

D. Perugini, *The Mixing of Magmas*, Advances in Volcanology,
https://doi.org/10.1007/978-3-030-81811-1

outcrops and the relative compositional variations. This paved a new ground for a better understanding of the fundamental physical and chemical mechanisms governing magma mixing.

The development of chaotic dynamics in magma bodies can also have a strong impact on mineral phases, which can act as recorders of the chemical and physical disequilibrium caused by magma mixing. The response of minerals to the modulation of disequilibrium by chaotic flow fields depends on the nature of the flow in which they are trapped in. If the crystals are located within chaotic regions they will be strongly subjected to the physical and chemical disequilibrium. On the contrary, if the crystals are incorporated within regular regions they might preserve their original identity because the closed structure of the flow fields will act as a barrier to disequilibrium. In any case, the compositional zoning in minerals represents an extraordinary archive of information to decipher the space and time evolution of magma bodies.

The fact that the presence of chaotic dynamics is well evident in igneous rocks that recorded mixing processes between compositionally different magmas is essentially due to the fact that the magmas act as fluid-dynamic markers of the process. This does not mean, however, that the presence of chaotic dynamics has to be restricted to magma mixing systems. Chaotic flow fields can develop in any fluid dynamic system. They can also originate within a single magma batch, which can, therefore, exhibit large degrees of spatial and temporal inhomogeneity. This is a key point opening new intriguing questions about, for example, the most appropriate geochemical models to be used to study the compositional variability of igneous rocks. Are the conventional mass balance geochemical models appropriate to capture the complexity inherent to magma evolutionary processes? I'll leave the possible answers to the imagination of the reader.

From the arguments presented in this book it emerges that chaotic dynamics and the resulting fractal patterns constantly permeate the evolution of magmatic systems. Their evolution can be seen as a non-linear cascade of events, which starting from the microscale are progressively amplified determining the behavior of these systems at the macroscale. Chaotic dynamics and fractals can be, hence, considered "universal" processes and patterns whose propagation in space and time governs the evolution of magmatic systems.

Finding the right way to face the complexity inherent to igneous systems is not easy. It is like trying to reconstruct a mosaic starting from a huge number of tiles initially placed randomly that need to be ordered. This is how Nature initially appears to us.

Considering the concepts of Chaos Theory and Fractals in the reconstruction of the mosaic allows us not to determine a priori the way in which the tiles will be positioned. This is a fundamental point. Chaos Theory informs us that it is not possible to know with infinite precision all the initial conditions of a dynamic system and, therefore, we cannot predict its long-term evolution. It is not useful to ignore this in the study of petrological and volcanological processes, as we might benefit from very powerful concepts and techniques to negotiate directly with the natural world that certainly will not be subjugated by the deterministic load that we assigned to it for centuries.

Chaos Theory and Fractal Geometry pose no limits to Nature, who can get rid of this deterministic load and express all its creativity, modulating its particular cases on a universal basis, rather than repeating itself continuously with predetermined and predictable patterns.

Perhaps we are barely scratching the surface of a new world. The exploration of this world may open new and unexplored fields of research and provide information about additional processes that can play a fundamental role in modulating the compositional variability of igneous rocks. This is a very challenging task requiring further considerable theoretical, numerical, and experimental efforts. Nevertheless, it might be worthwhile trying since a deeper understanding of igneous systems and their role in planetary differentiation and volcanism will force us to face, sooner or later, the process of mixing and its consequences.

www.ingramcontent.com/pod-product-compliance
Ingram Content Group UK Ltd.
Pitfield, Milton Keynes, MK11 3LW, UK
UKHW051927230425
457752UK00004B/20
* 9 7 8 3 0 3 0 8 1 8 1 0 4 *